Governors State University
Library
Hours:
Monday thru Thursday 8:30 to 10:30
Friday and Saturday 8:30 to 5:00
Sunday 1:00 to 5:00 (Fall and Winter Trimester Only)

DEMCO

Chemistry of Natural Products

Amino acids, Peptides, Proteins and Enzymes

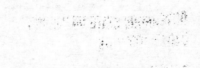

Chemistry of
Natural Products

Amino acids, Peptides, Proteins and Enzymes

V. K. Ahluwalia
Lalita S. Kumar
Sanjiv Kumar

Taylor & Francis
Taylor & Francis Group
Boca Raton London New York

CRC is an imprint of the Taylor & Francis Group,
an informa business

Ane Books India

© Copyright 2006 Ane Books India

Reprinted in 2007 by
Ane Books India

4821 Parwana Bhawan, 1st Floor
24 Ansari Road, Darya Ganj, New Delhi -110 002, India
Tel: +91 (011) 2327 6843-44, 2324 6385
Fax: +91 (011) 2327 6863
e-mail: anebooks@vsnl.com
Website: www.anebooks.com

For

CRC Press
Taylor & Francis Group
6000 Broken Sound Parkway, NW, Suite 300
Boca Raton, FL 33487 U.S.A.
Tel : 561 998 2541
Fax : 561 997 7249 or 561 998 2559
Web : www.taylorandfrancis.com

For distribution in rest of the world other than the Indian sub-continent

ISBN-10 : 1 42005 917 3
ISBN-13 : 978 1 42005 917 5

British Library Cataloguing in Publication Data
A catalogue record for this book is available from the British Library

Printed at Gopsons Paper Ltd, Noida (U.P.), India

PREFACE

The study of 'Natural Products' is an integral part of the undergraduate and the postgraduate curriculum across universities in India. A need for a comprehensive text covering different areas of natural product chemistry in reasonable detail is felt by the students and teachers alike. This book offers readers just the right information in a most concise and articulate manner without diluting the importance of the subject matter.

The book is designed for the undergraduate (both at pass and honours level) and the postgraduate students of Chemistry. The text would be equally useful to the learners in interdisciplinary areas like Biochemistry, Biotechnology etc. and would provide a good foundation to the ones venturing into research in the areas covered in the text especially peptide synthesis.

The text is divided into four chapters – a chapter each dedicated to amino acids, peptides, proteins and enzymes. The first chapter on 'Amino Acids' covers nomenclature, classification, stereochemistry, physical and chemical properties, synthesis and industrial applications. The important reactions have been explained with the help of mechanisms involved.

The second chapter on 'Peptides' explains the formation and structure of peptide bond and its significance. The chapter gives a detailed account of the solution phase and solid phase synthesis of peptides besides discussing the structure and functions of some biologically important peptides.

The third chapter on 'Proteins' discusses different ways of classification and explains their properties. The structural organisation of proteins has been covered in adequate details including the primary structure determination.

The fourth and the last chapter on 'Enzymes' is a logical extension of the coverage in the first three chapters. It covers the classification, nomenclature and mode of action of enzymes besides a detailed account of the structure and function of different coenzymes.

The authors wish to acknowledge Prof. J.M. Khurana of university of Delhi to have gone through the manuscript. We look forward to critical comments from the users of the text and would welcome pointers to errors, omissions and obscurities (if any) so as to improve the book for future use.

Lastly, the authors express their sincere thanks to Mr. Sunil Sexena and all staff members of Ane Books India, for their help and the excellent cooperation received.

V. K. Ahluwalia
Lalita S. Kumar
Sanjiv Kumar

CONTENTS

Chapter 1

Amino Acids

1.1 INTRODUCTION

A large number of amino acids are found in nature, most of which are α-amino acids. These occur in free form or as constituents of other biomolecules like, peptides, proteins, coenzymes, hormones etc. Of these, twenty[*] α-amino acids (in fact, nineteen amino and one imino acid) are found to be fundamental to the sustenance of the life forms. These are utilised in the synthesis of peptides and proteins under genetic control, which in turn are vital for life. These α-amino acids are referred to as **coded amino acids** or **primary protein amino acids** or **proteinogenic amino acids.** The rest of the amino acid are referred to as **non-coded amino acids** or **non-protein amino acids**. These are usually formed by post - translational modification, i.e., modified after translation (protein biosynthesis). These modifications are generally essential for the function of the protein. At the present level, we would confine ourselves primarily to the coded or protein amino acids which are also called 'standard' or 'canonical' α-amino acids. Though the coded amino acids primarily serve as the building blocks of proteins, these are precursors to many other important compounds. For example, tryptophan is a precursor of the neurotransmitter serotonin, glycine is one of the reactants in the synthesis of haeme–a porphyrin while serine is a constituent of phosphatidyl serine–a phosphoglyceride and so on.

The presence of an acid and amino functional groups in the same molecule make the α-amino acids show interesting acid-base and other physical properties. These properties in turn are crucial to the structure and the biological roles played by α-amino acids and of the molecules derived from them. In addition to the α-amino acids many amino acids with amino

---

[*] In addition to these, two more amino acids by the names selenocysteine and pyrrolysine have been reported to be found in proteins. These amino acids do have genetic codes (UGA and UAG respectively) but are not incorporated into the proteins in the usual way. Selenocysteine is found in several enzymes like *glutathione peroxidase, formate dehydrogenase* etc., while pyrrolysine is found in microbes that produce methane.

group at β or γ positions or even more separated from the carboxyl group are also known and have important biological roles.

1.2 NOMENCLATURE OF AMINO ACIDS

The amino acids obtained by hydrolysis of proteins with boiling aqueous acid or base, contain an amino as well as a carboxyl group (the amino group being α to the carboxyl group) and are generally known as α-amino acids. In general, an α-amino acid consists of an amino group, a carboxyl group, a hydrogen atom and a distinctive R group also called a side chain, bonded to a carbon atom, which is called the α-carbon. The general formula of an α-amino acid can be represented as shown below. The meaning and the significance of the charged structure are discussed in Sec. 1.5.2.

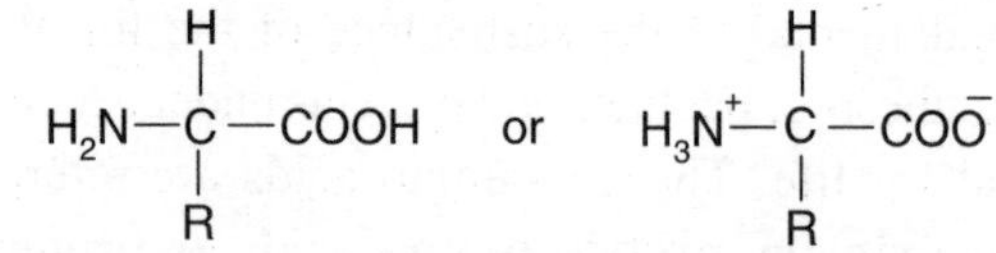

$$H_2N-\overset{\overset{\displaystyle H}{|}}{\underset{\underset{\displaystyle R}{|}}{C}}-COOH \quad \text{or} \quad H_3\overset{+}{N}-\overset{\overset{\displaystyle H}{|}}{\underset{\underset{\displaystyle R}{|}}{C}}-CO\bar{O}$$

General structure of an α-amino acid

The traditional and well known names of the common α-amino acids were, generally, given to them by their discoverers and bear no relationship to their chemical structures. These α-amino acids were normally named either on the basis of the source from which they were obtained or on some special property associated with them. For example, glycine derived its name from its sweet taste (Greek *glycos*, meaning sweet) while tyrosine got its name from its source namely, the milk protein, casein (Greek. *tyros*, meaning cheese). The semi-systematic names of substituted α-amino acids are formed according to the general principles of organic nomenclature, by attaching the name of the substituent group to the trivial name of the amino acid. For example, N-methylglycine. A list of the coded α-amino acids along with the structures of their side chains, IUPAC names, one and three letter abbreviations (Sec. 1.2.1) is given in Table 1.1.

According to IUPAC, an α-amino acid that is otherwise similar to one of the common ones but contains one more methylene group in the carbon chain, may be named by prefixing 'homo' to the name of that common amino acid. For example, homoserine and homocysteine are higher homologs of serine and cysteine respectively.

Table 1.1 Names, symbols and structural formulae of coded α-amino acids

Common name	Three letter symbol	One letter symbol	IUPAC name	Structural formula	Frequency of occurrance[‡] (%)
Alanine	Ala	A	2-Aminopropanoic acid	$CH_3-CH(NH_2)-COOH$	7.7
Arginine[**]	Arg	R	2-Amino-5-guanidinopentanoic acid	$H_2N-C(=NH)-NH-(CH_2)_3-CH(NH_2)-COOH$	5.1
Asparagine	Asn	N	2-Amino-3-carbamoylpropanoic acid	$H_2N-CO-CH_2-CH(NH_2)-COOH$	4.3
Aspartic acid	Asp	D	2-Aminobutanedioic acid	$HOOC-CH_2-CH(NH_2)-COOH$	5.2
Cysteine	Cys	C	2-Amino-3-mercaptopropanoic acid	$HS-CH_2-CH(NH_2)-COOH$	2.0
Selenocysteine	Sec[†]	U[†]	2-amino-3-selenopropanoic acid	$HSe-CH_2-CH(NH_2)-COOH$	—
Glutamine	Gln	Q	2-Amino-4-carbamoylbutanoic acid	$H_2N-CO-(CH_2)_2- CH(NH_2)-COOH$	4.1
Glutamic acid	Glu	E	2-Aminopentanedioic acid	$HOOC-(CH_2)_2- CH(NH_2)-COOH$	6.2
Glycine	Gly	G	Aminoethanoic acid	$CH_2(NH_2)-COOH$	7.4
Histidine[**]	His	H	2-Amino-3-(1H-imidazol-4-yl)-propanoic acid	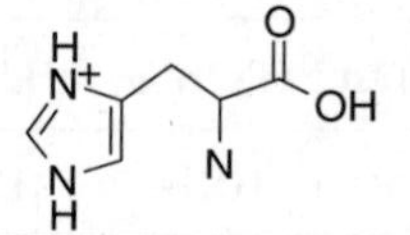	2.3
Isoleucine[*]	Ile	I	2-Amino-3-methylpentanoic acid	$C_2H_5-CH(CH_3)-CH(NH_2)-COOH$	5.3
Leucine[*]	Leu	L	2-Amino-4-methylpentanoic acid	$(CH_3)_2CH-CH_2-CH(NH_2)-COOH$	8.5

Common name	Three letter symbol	One letter symbol	IUPAC name	Structural formula	Frequency of occurr-ance[‡] (%)
Lysine[*]	Lys	K	2,6-Diaminohexanoic acid	H_2N-$(CH_2)_4$-$CH(NH_2)$-COOH	5.9
Methionine[*]	Met	M	2-Amino-4-(methylthio) butanoic acid	CH_3-S-$(CH_2)_2$- $CH(NH_2)$-COOH	2.4
Phenylalanine[*]	Phe	F	2-Amino-3-phenylpropanoic acid	C_6H_5-CH_2-$CH(NH_2)$-COOH	4.0
Proline	Pro	P	Pyrrolidine-2-carboxylic acid	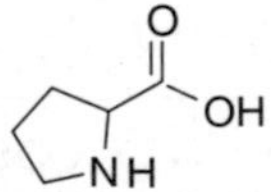	5.1
Serine	Ser	S	2-Amino-3-hydroxypropanoic acid	HO-CH_2-$CH(NH_2)$-COOH	6.9
Threonine[*]	Thr	T	2-Amino-3-hydroxybutanoic acid	CH_3-$CH(OH)$-$CH(NH_2)$-COOH	5.9
Tryptophan[*]	Trp	W	2-Amino-3-(IH-indol-3-yl)-propanoic acid	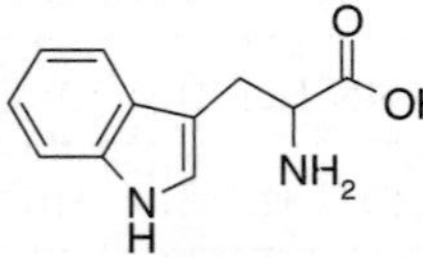	1.4
Tyrosine	Tyr	Y	2-Amino-3-(4-hydroxyphenyl)-propanoic acid	HO-C_6H_4-CH_2-$CH(NH_2)$-COOH	3.2
Valine[*]	Val	V	2-Amino-3-methylbutanoic acid	$(CH_3)_2CH$-$CH(NH_2)$-COOH	6.6

* Essential amino acids   ** Semi-essential amino acids

† These are suggested symbols; not officially accepted as yet.

‡ From Jones, D.T. Taylor, W.R. & Thornton, J.M. (1991) *CABIOS* 8, 275-282.

1.2.1 Representation of Amino Acids

The α-amino acids are commonly represented in terms of *three letter symbols*. In this convention, the symbol for an amino acid is derived from its trivial name and includes the first three letters of this name (asparagine, glutamine, isoleucine and tryptophan being exceptions). The three letter symbols are written as one capital letter followed by two lowercase letters, e.g. glutamic acid is represented as Glu (not GLU or glu), regardless of its position in a sentence. The three letter symbols of coded amino acids are given in Table 1.1. These are used primarily to save space while representing peptides (obtained by condensation of two or more amino acids) and actually represent the unsubstituted amino acids. However, to represent the derivatives of the amino acids or amino acid residues in a peptide or protein, the symbols are modified by hyphens. The hyphen before the symbol of the amino acid indicates a bond formation with the α-nitrogen while the one after the symbol of the amino acid is indicative of the bond formation to the carboxyl group. For example, Ac–Ala represents the N-acetyl derivative of alanine while Ala–OMe stands for the methyl ester of alanine. –Ala– represents an alanine residue in the peptide (Chapter 2). For example, Gly-Ala-Val represents the tripeptide glycylalanylvaline in which alanine is the central residue. Further, as a convention, these amino acid symbols denote the L- configuration of chiral amino acids (explained later) unless otherwise indicated by the presence of D- or DL- before the symbol and separated from it with a hyphen for example D-Ala.

Nowadays *one letter system* is used more often. In this system each α-amino acid has been assigned a one letter symbol and the peptide or the protein is represented as a string of letters starting with the N-terminal (Sec. 2.3) amino acid moving towards C-terminal. The one letter symbol for eleven of the twenty coded amino acids is the first letter of their full name and for the rest nine distinct letters have been assigned so as to avoid any ambiguity (Table 1.1). The three letter abbreviations are quite straightforward, however, the relationship of one letter abbreviations to the names of the amino acids is somewhat less obvious. These are commonly used in representing long sequences as these save space and are less likely to be confused. For example, Q, E and G, the one letter symbols of glutamine, glutamic acid and glycine respectively are less likely to be confused than the obvious three letter symbols viz., Gln, Glu and Gly respectively. The three letter system, on the other hand has advantages over one letter system as it has provision for representing the protecting groups and other structural details of the amino acids e.g., Boc-Gly represents the

N-tertiarybutyloxocarbonyl derivative of glycine. Further, one letter symbols are **never** used to represent the individual amino acids in other contexts. A pentapeptide, glycylvalyltyrosylprolylglycine would be represented as Gly–Val–Tyr–Pro–Gly and GVYPG respectively in the three letter and one letter conventions. It may be noted here that while writing the name of the peptide, the suffix **–ine** of the amino acid name is replaced by **–yl**. This aspect has been discussed later in detail (Chapter 2).

In representing the amino acid sequences of peptides and proteins as derived from amino acid analysis (Sec. 3.4.1), two additional symbols viz. Glx and Asx are used. These ambiguous symbols are used due to uncertainties in the determination of glutamine and asparagine. In the course of their determination (acidic / basic conditions) the Gln and Asn residues may hydrolyse to give Glu and Asp respectively. It is difficult to ascertain whether the Glu (or Asp) residue determined was originally a Glu (or Asp) or it was obtained from Gln (or Asn). In terms of one letter symbols Glx and Asx residues are represented as 'X'.

Different atoms of the side chains of the amino acids are usually indicated in terms of Greek alphabets as shown for the amino acid, lysine in Fig 1.1. It may be noticed that lysine contains an ε-amino group.

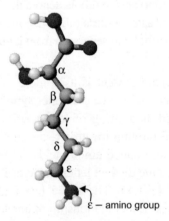

Fig. 1.1 *Structure of lysine showing Greek designations of the side chain carbon atoms.*

The nomenclature of α-amino acids and the conventions of representing them pertain to the ***coded amino acids***. However, the names of some ***non-coded amino acids*** which are closely related to coded amino acids are based on the three letter symbols of the coded amino acids e.g., L-Hypro represents *trans*-4-hydroxy -L-proline that is obtained by post translational hydroxylation of proline residue. As could be seen by now that these conventions are used for simplifying the representations of the structures of peptides and proteins. The systematic names and formulae given in Table 1.1 refer to hypothetical forms in which amino groups are unprotonated and carboxyl groups are undissociated. It should not be taken to imply that these structures represent an appreciable fraction of the amino acid molecules under physiological conditions.

1.3 CLASSIFICATION OF AMINO ACIDS

In a broad sense amino acids can be put into two groups–the *coded amino acids* and the *non-coded amino acids*. In the process of protein biosynthesis the coded amino acids or the *primary protein amino acids* are incorporated one by one as per the instructions contained in the 'gene' for the concerned protein. The process is referred to as **translation** as it entails the translation of the genetic code to the amino acid sequence. Once the protein is synthesised, some modifications referred to as **post translational modifications** can occur in some amino acid residues of the protein. Thus proteins may contain certain amino acids other than the twenty coded amino acids. These are called as **secondary protein amino acids**. If however, the modification leads to cross linking of two amino acids then these are referred to as **tertiary protein amino acids**. As mentioned earlier we would confine ourselves primarily to the coded or primary protein amino acids.

1.3.1 Coded or Primary Protein Amino Acids

The coded or primary protein amino acids with the general structure given earlier (Sec.1.2) contain different side chains varying in size, shape, charge, hydrogen bonding capacity and chemical reactivity. On the basis of structures and nature of the side chains, these amino acids are classified broadly into three groups–nonpolar (or apolar), neutral polar and charged polar amino acids.

1.3.1.1 Nonpolar or Apolar Amino Acids

As many as ten coded amino acids belong to this class. These include glycine, alanine, valine, isoleucine, leucine, proline, cysteine, methionine, tryptophan and phenylalanine. These amino acids are usually located on the interior of the protein as their side chains are hydrophobic in nature. Glycine, alanine, valine, leucine, isoleucine and proline side chains are purely aliphatic. Of these, proline is unique in the sense that it contains an aliphatic side chain that is covalently bonded to the nitrogen atom of the α-amino group, forming an imide bond and leading to a constrained 5-membered ring. The side chains (as shown below) in the case of valine, leucine and isoleucine are bifurcated. This bifurcation or branching is close to the main chain and can restrict the conformation of the polypeptide by steric hindrance.

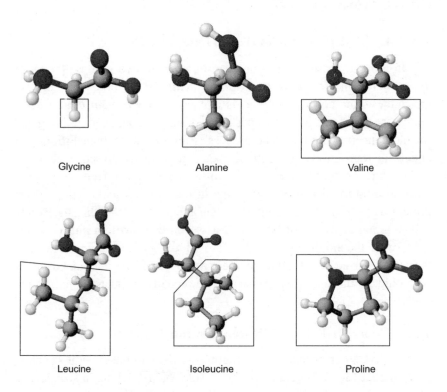

Glycine   Alanine   Valine

Leucine   Isoleucine   Proline

    The nonpolar side chains generally have low solubility in water because they can have only van der Waals interactions with water molecules. However, tryptophan with a nitrogen atom in its indole ring system is somewhat polar and can have hydrogen bonding interactions with other residues or even solvent molecules. The aromatic ring of phenylalanine is quite hydrophobic and chemically reactive only under extreme conditions, though its ring electrons are readily polarised. The sulphydryl (thiol) group of cysteine can ionise at slightly alkaline pH and can react with a second sulphydryl group to form a disulfide bond. On the other hand, methionine has a long alkyl side chain containing a sulphur atom that is relatively inert as a hydrogen bond acceptor.

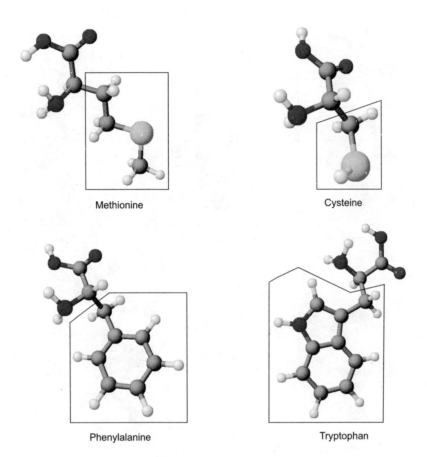

Methionine

Cysteine

Phenylalanine

Tryptophan

1.3.1.2 Neutral or Uncharged Polar Amino Acids

The amino acids belonging to this category have side chains with an affinity for water but are not charged. Serine, threonine, tyrosine, aspargine and glutamine belong to this category. Of these the first three amino acids have a hydroxyl group while the latter two have amide groups in their side chain. Due to the presence of hydroxyl groups, serine, threonine and tyrosine can function as both donors and acceptors in hydrogen bond formation. The phenyl ring of tyrosine permits stabilisation of the anionic phenolate form obtained on the ionisation of the hydroxyl proton, which has a pK_a of about 10. Serine and threonine however cannot be deprotonated at ordinary pH. Due to the presence of amide group, asparagine and glutamine side chains are relatively polar and can act as donors or acceptors in the formation of hydrogen bonds.

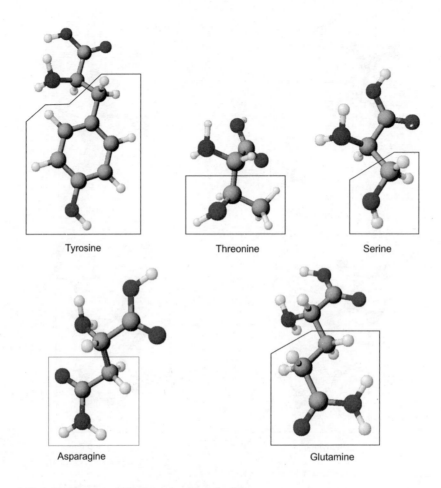

Tyrosine   Threonine   Serine

Asparagine   Glutamine

1.3.1.3 Charged Polar Amino Acids

As the name suggests, the members of this group viz., lysine, arginine, histidine, glutamic acid and aspartic acid have polar side chains due to the presence of positive or negative charge at the end of their side chain. The lysine ε-amino group has a pK_a value close to 10, while the guanidino group in the arginine side chain has a pK_a value of ~12. Histidine is another basic residue with its side chain organised into a closed ring structure that contains two nitrogen atoms. Glutamic and aspartic acid are two amino acids that contain a carboxyl group in their side chain. These differ only in the number of methylene groups in the side chain, with one and two methylene groups, respectively. Their carboxylate groups are extremely polar and can readily form hydrogen bonds, by acting as a donor or acceptor. These have pK_a values of about 4.5.

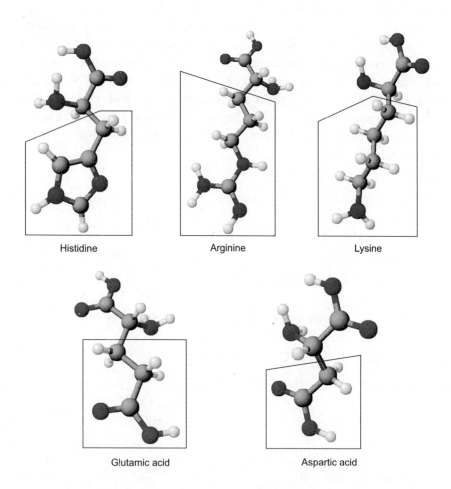

Histidine   Arginine   Lysine

Glutamic acid   Aspartic acid

1.3.2 Secondary and Tertiary Protein Amino Acids

In addition to the twenty α-amino acids that are the primary components of proteins, some other amino acids exist naturally. These are obtained by modification of the primary protein amino acids. The common modifications include simple derivatisation like, hydroxylation, methylation, acetylation, carboxylation and phosphorylation on the side chains of some amino acids. These lead to **secondary protein amino acids or secondary amino acids.** For example, hydroxylysine and hydroxyproline are simply functionalised derivatives of primary protein amino acids, lysine and proline respectively. These are found in collagen, a common structural protein.

Some secondary amino acids are obtained by replacing the hydrogen atom of OH, SH or NH group in the side chain of the primary protein amino acid by glycosyl, phosphate or a sulphate group. In some cases, larger groups like polymeric carbohydrates or lipids are attached to amino acid residues. Such secondary amino acids are commonly found in glycoproteins. In some other cases the primary protein amino acids are derivatised at the N-terminal as a methyl derivative to form a secondary amino acid. For example, secondary amino acid like, N^{ω}-methylarginine is commonly found in histones, the nuclear proteins. The structures of some secondary protein amino acids are given in Fig. 1.2.

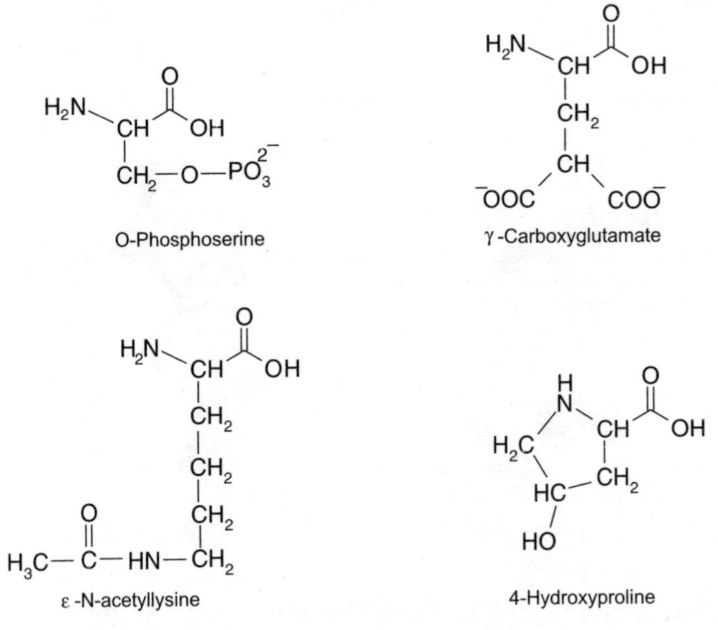

O-Phosphoserine

γ-Carboxyglutamate

ε-N-acetyllysine

4-Hydroxyproline

Fig. 1.2 *Some secondary protein amino acids.*

Crosslinking of amino acids in a protein generates **tertiary protein amino acids**. A disulphide bridge formed by the oxidation of the thiol groups of two cysteine residues to give cystine is the most common cross-link occurring in proteins. Sometimes the side chain carboxyl group of a glutamic acid residue can get into a cross-link with the ε- amino group of lysine to give a tertiary protein amino acid, ε-(γ-glutamyl) lysine. The structures of some tertiary protein amino acids are given in Fig.1.3.

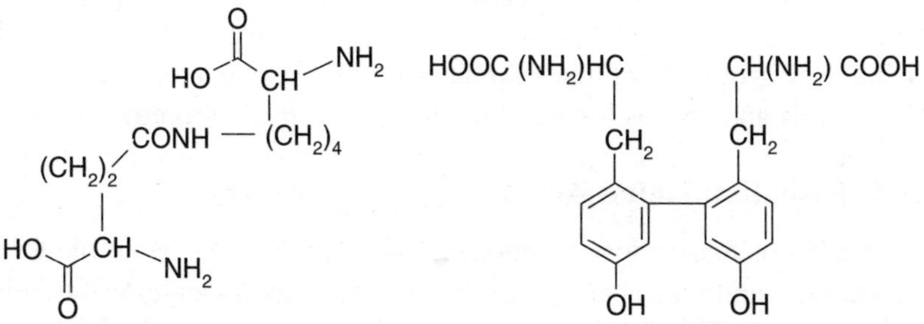

ε-(γ-glutamyl) Lysine 3, 3'-Bityrosine

Fig. 1.3 *Some tertiary protein amino acids.*

1.3.3 Non-Coded or Non-Protein Amino Acids

Non-protein amino acids (also called non-standard amino acids) are those amino acids which are neither found in proteins assembled during protein biosynthesis nor are generated by post-translational modifications. This may be due to the lack of a specific codon (genetic code) and t-RNA. Hundreds of such amino acids are known and a large number of these are α-amino acids. These non-coded amino acids are found mostly in plants and microorganisms and arise as intermediates or as the end product of the metabolic pathways. For example, N-acetyltyrosine is formed in the metabolism of tyrosine; betaine is involved in glycine biosynthesis and citrulline participates in ornithine cycle. These may also arise in the process of detoxification of the compounds of foreign origin. The structures of some non-protein amino acids are given in Fig.1.4.

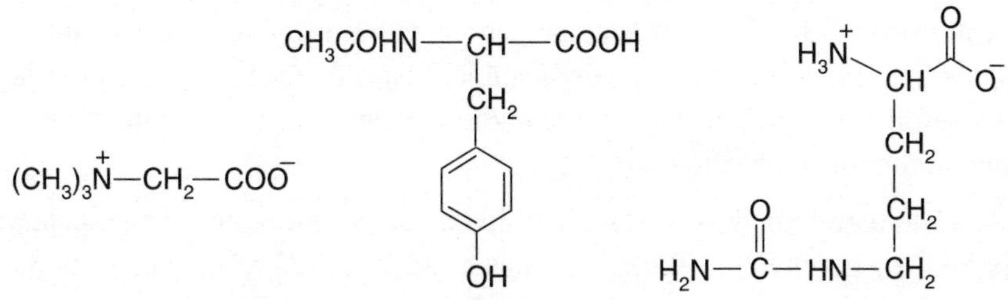

Betaine N-Acetyltyrosine Citrulline

Fig. 1.4 *Some non-protein amino acids.*

It is difficult to ascribe an obvious direct function to most of these amino acids in an organism. However, most of the functions of these amino acids in plants and microorganisms may be associated with other organisms in the environment. Canavanine–a homologue of arginine in which the δ-methylene group of arginine is replaced by an oxygen atom, is found in alfalfa seeds and acts as a natural defense against insect predators.

1.3.4 Essential Amino Acids

In addition to the above mentioned classification based on the nature of side chains, amino acids are also classified as essential and non-essential amino acids on the basis of their source in the living system. Of the twenty standard amino acids listed in Table 1.1, in case of humans, more than half of these can be made by the body itself, while the others, called **essential amino acids**, must come from the diet. These are required to maintain the nitrogen balance in the body. In fact, the meaning of the term essential differs from one species to the other. The classification of an amino acid as essential or non-essential does not reflect its importance because all the twenty amino acids are necessary for normal functioning of the body. This classification system simply reflects whether or not the body is capable of manufacturing a particular amino acid. The essential amino acids (for human beings) are isoleucine, leucine, valine, lysine, methionine, phenylalanine, threonine and tryptophan.

Some other amino acids have also been added to this list. These are the ones that are synthesised in the body but not at a rate required for the normal growth of the organism. The amino acids, arginine and histidine, belong to this list of **semi-essential** amino acids because the body does not always require dietary sources for it. Two amino acids viz., cysteine and tyrosine occupy ambiguous position in this classification. These are synthesised in the body in adequate amounts but use two essential amino acids viz., methionine and phenylalanine respectively for their synthesis. Therefore, their presence in the diet indirectly decreases the requirement of methionine and phenylalanine.

The requirement of essential amino acids per kilogram of the dietary protein are given in the Table 1.2. This is called the reference pattern of the amino acids and acts as a standard to determine the quality of the protein being consumed. Further, a number of essential amino acids are toxic if taken in excess. For example, a diet containing excess of leucine may cause pellagra. Therefore, a caution should be exercised while taking protein or amino acid supplements.

Table 1.2 The requirement of essential amino acids

Essential amino acid	Requirement (g per kg dietary protein)
Isoleucine	42
Leucine	48
Lysine	42
Methionine	22
Phenylalanine	28
Threonine	28
Tryptophan	14
Valine	42

1.4 STEREOCHEMICAL ASPECTS OF α-AMINO ACIDS

The tetrahedral array of four different groups about the α-carbon atom of α-amino acids (with the exception of glycine) makes it asymmetric–a chiral carbon. This asymmetry produces optical or stereoiomers in the amino acids. It is known that different amino acids contain different side chains (R groups) varying in size, shape, charge, hydrogen bonding capacity and chemical reactivity. Glycine is the simplest amino acid with just one hydrogen atom as its side chain. Due to the absence of a chiral atom glycine is not optically active.

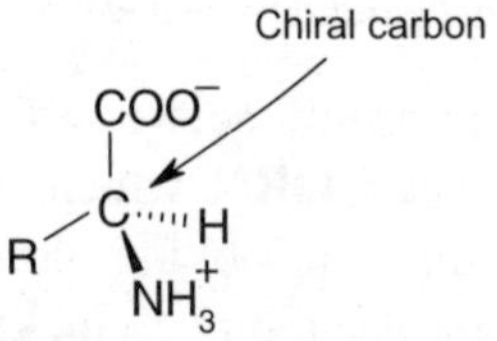

It is advisable to draw the structures of amino acids in the **Fischer-Rosanoff** convention. For this, the chiral atom of the amino acid is projected onto the plane of the paper in such a way that the central atom appears as the point of intersection of a vertical line that joins three atoms of the principal chain i.e., $COO^- - C_\alpha - R$ and a horizontal straight line joining the attached groups viz., NH_3^+ and H. The central atom is considered to lie in the plane of the paper, the atoms of the principal chain behind the plane from the viewer, and the remaining two groups in front of the plane. The Fischer-Rosanoff representation of an amino acid may be drawn as shown in Fig. 1.5.

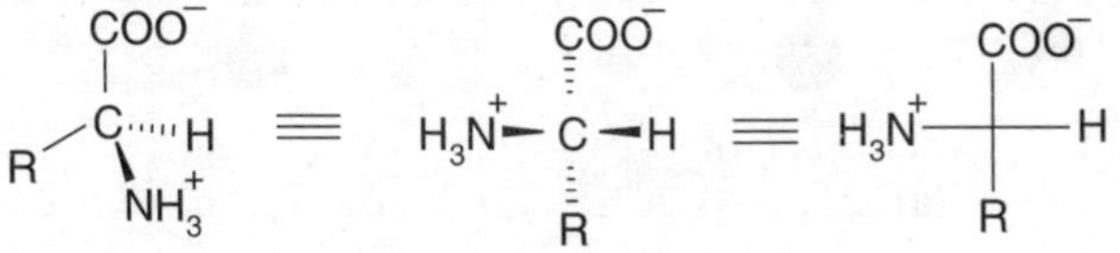

Fig. 1.5 *The Fischer-Rosanoff representation of an α-amino acid.*

1.4.1 Absolute Configuration of α-Amino Acids

The absolute configuration at the chiral carbon atom of α-amino acids is designated by a small size capital letter prefix, D or L. This prefix indicates a formal relationship of the given amino acid to D- or L-serine–a standard used for correlating the configurations of the amino acids. The configurations of D- or L-serine, on the other hand are related to D- or L-glyceraldehyde respectively. In other words, correlation of the absolute configuration of the amino acid with D- or L-serine is equivalent to their correlation with D- or L-glyceraldehyde. It is reemphasised here that these D-and L-notations do not indicate anything about the optical rotation of the amino acids. The relationship between the two isomeric forms of serine and the corresponding glyceraldehydes is shown in Fig. 1.6. All the natural α-amino acids are found to be L-amino acids.

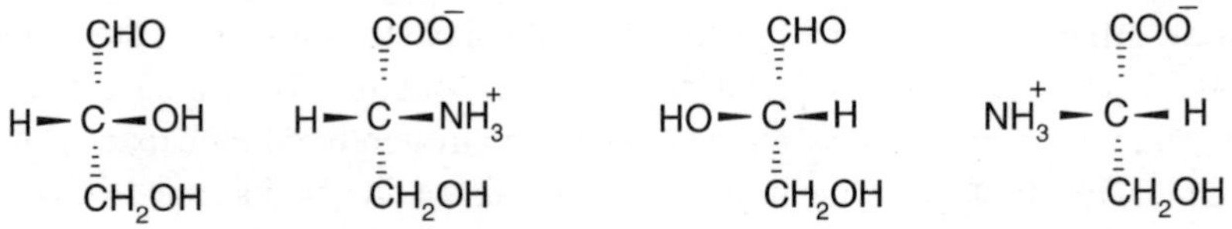

CHO	COO⁻	CHO	COO⁻
H—C—OH	H—C—NH₃⁺	HO—C—H	NH₃⁺—C—H
CH₂OH	CH₂OH	CH₂OH	CH₂OH
D-Glyceraldehyde	D-Serine	L-Glyceraldehyde	L-Serine

Fig. 1.6 *Correlation of absolute configuration of D- and L-serine with D- and L-glyceraldehyde.*

To ascertain whether an amino acid is a L-amino acid or a D-amino acid we may use the mnemonic **CORN**, which is an acronym for **–COOH**, **–R** and **–NH₂** groups. For this, on looking along the hydrogen-α-carbon bond of an amino acid, if the eyes move **clockwise** as we move through –COOH group, then the –R group and then the –NH₂ group starting at the carboxylic acid group (Fig. 1.7), then the amino acid has L-configuration else it belongs to the D-configuration.

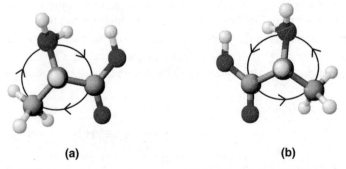

(a)   **(b)**

Fig. 1.7 *Three dimensional structures of (a) L-alanine and (b) D-alanine.*

1.4.1.1 The *RS* Notation

In the more general system of stereochemical designation, i.e., the *RS* or *CIP* (Cahn–Ingold–Prelog) system, the ligands of a chiral atom are placed in an order of preference, according to an arbitrary priority scheme. In this scheme the atom of higher atomic number takes precedence over that of the lower atomic number (e.g., –OH precedes –NH$_2$). However, if the first atom of the substituent group happens to be same then the preference is decided on the basis of the next atom (e.g., –CH$_2$OH precedes –CH$_3$). These groups are labelled as 1,2,3 and 4 (or W, X, Y and Z). On viewing from the side opposite to the least-preferred (4th or Z) group, if the first three groups appear in clockwise order the chiral centre is designated as *R* (from Latin, *rectus*); if anticlockwise, it is *S* (from Latin, *sinister*), Fig. 1.8.

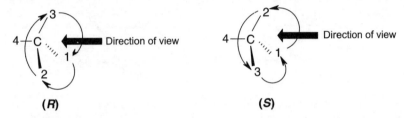

(R)   **(S)**

Fig. 1.8 *The relative disposition of different substituents around the chiral atom in R and S notations. The numbers indicate the order of preference.*

The L-configuration, possessed by the chiral α-amino acids found in proteins, nearly always corresponds to *S* or sinister notation, because in most amino acids the order of preference of the groups around the chiral carbon atom is NH$_3^+$ (1), COO$^-$ (2), R (3) and H (4), Fig. 1.9 (a). However, the amino acids L-cysteine and L-cystine are exceptions as these have *R* notation. It is due to the presence of a sulphur atom in the side chain (i.e., – CH$_2$SH). Here, since the atomic number of sulphur is higher than that of oxygen, the group *R* takes precedence over carboxylate ion in these amino acids, Fig. 1.9(b). The order is NH$_3^+$ (1), CH$_2$SH (2), COO$^-$ (3) and H (4).

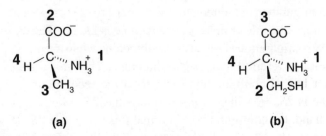

(a)   **(b)**

Fig. 1.9 *The order of preference of substituents in (a) alanine and (b) cysteine.*

A mixture of equimolar amounts of D-and L-amino acids containing one chiral centre is termed racemic and is designated by the prefix DL (no comma between D and L), e.g. DL-leucine or it may be designated by the prefix *rac-* (e.g. *rac-*leucine).

1.4.2 Amino Acids with Two Chiral Centres

Historically, amino acids with two chiral centres were named by allotting a name (common or trivial) to the first diastereoisomer to be discovered. The second diastereoisomer, when found or synthesised, was then assigned the same name but with the prefix allo-. Since this method can be used only with trivial names it has now been recommended that the prefix allo- should be used only for the amino acids, isoleucine and threonine.

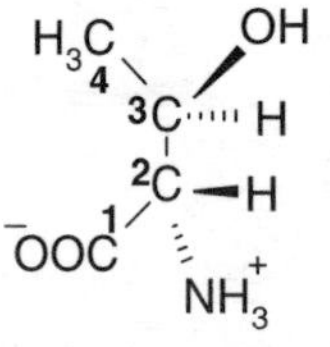

2S,3R-Threonine
(L-Threonine)

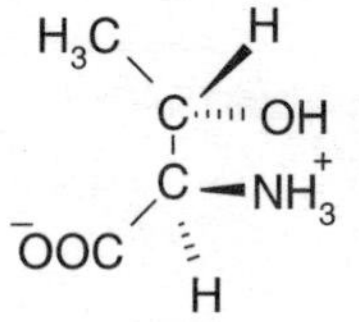

2R,3S-Threonine
(D-Threonine)

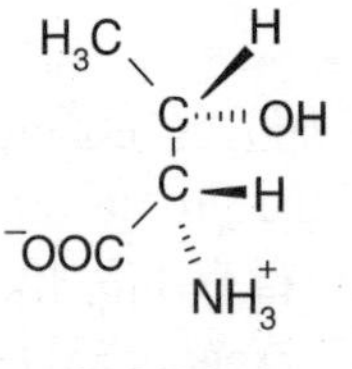

2S,3S-Threonine
(L-*allo*-Threonine)

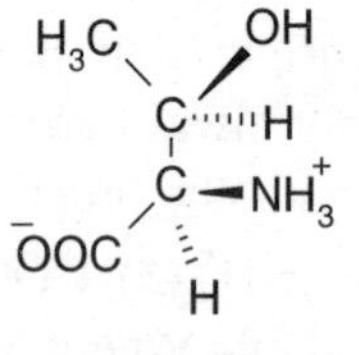

2R,3R-Threonine
(D-*allo*-Threonine)

In amino acids with two chiral centres, the D- and L-notations refer to the configuration of the α-carbon atom. Threonine and isoleucine both have a second asymmetric center at carbon position 3 along their chains. The absolute configuration of L-threonine, the normal protein component, is 2S,3R. Its enantiomer (the mirror image) , D-threonine is 2R, 3S. The diastereomer with 2S, 3S configuration is called L-*allo*-threonine while the one with 2R,3R configuration is referred to as D-*allo*-threonine. The *allo-* form of a given enantiomer has its configuration inverted at the second chiral atom e.g., L-threonine is 2S, 3R while L-*allo*-threonine has 2S, 3S configuration. On the other hand, the configuration of normal L-isoleucine is 2S, 3S while its enantiomer is 2R, 3R. The diastereomers (2R, 3S and 2S, 3R) are called D-*allo*-isoleucine and L-*allo*-isoleucine respectively.

$$CH_3\text{-}CH_2 \diagdown \quad \diagup H$$
$$C^{\cdots\cdots}CH_3$$
$$C\text{---}H$$
$$\overline{O}OC \qquad NH_3^+$$

2S, 3S-Isoleucine

(L-Isoleucine)

$$CH_3\text{-}CH_2 \diagdown \quad \diagup CH_3$$
$$C^{\cdots\cdots}H$$
$$C\text{---}NH_3^+$$
$$\overline{O}OC \qquad H$$

2R, 3R-Isoleucine

(D-Isoleucine)

$$CH_3\text{-}CH_2 \diagdown \quad \diagup CH_3$$
$$C^{\cdots\cdots}H$$
$$C\text{---}H$$
$$\overline{O}OC \qquad NH_3^+$$

2S, 3R-Isoleucine

(L-allo-Isoleucine)

$$H_3C\text{-}CH_2 \diagdown \quad \diagup H$$
$$C^{\cdots\cdots}CH_3$$
$$C\text{---}NH_3^+$$
$$\overline{O}OC \qquad H$$

2R, 3S-Isoleucine

(D-allo-Isoleucine)

The optically inactive diastereoisomers of amino acids with two chiral centres are represented by a prefix *meso-* or *ms* if the optical inactivity is due to internal compensation e.g., *meso*-cystine, Fig. 1.10.

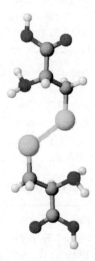

Fig. 1.10 *Three dimensional structure of meso-cystine.*

The direction of rotation of plane polarised light of specified wavelength in a specified solvent and temperature is denoted with a 'plus' or 'minus' sign in parenthesis. This may be done for emphasis, with or without a configurational symbol D or L, e.g. (+)-glutamic acid or (+)-L-glutamic acid. The sign of optical rotation precedes all other details of the compound like, substituents etc. e.g., (+)-6-hydroxytryptophan. A racemic mixture of amino acids, however, is indicated by (±), e.g. (±)-alanine.

1.5 PHYSICAL PROPERTIES OF α-AMINO ACIDS

The physicochemical properties of the amino acids are very important in determining the structure and function of the peptides and proteins derived from them. The presence of an amino group and an acid functional group in the same molecule along with the presence of variable side chains make amino acids show interesting properties especially the acid-base properties. Further the presence of a chiral carbon makes amino acids show optical properties.

1.5.1 General Physical Properties

Amino acids are colourless crystalline solids that are soluble in water, though to different extents. Cysteine, lysine and proline are highly soluble while alanine, arginine, glycine and threonine have moderate solubility and others have a low solubility (from 0.5 to ~ 9 g per 100 g). Tyrosine has an exceptionally low solubility of ~0.05 g per 100 g of water at 25 °C. None of the amino acids show a sharp melting point; all with the exception of cysteine and glutamine decomposes over 200 °C. The taste –a subjective property of the amino acids, varies from bitter to neutral to sweet. Cysteine however, has a sulphurous taste. The physical properties of twenty coded α-amino acids have been compiled in Table 1.3.

All the amino acids with the exception of glycine are optically active. The specific rotation ($[\alpha]_D^{25}$) values of the coded amino acids in water and 5M HCl are also given in Table 1.3. The $[\alpha]_D^{25}$ values in HCl are more positive or less negative than those in water. The optical rotatory dispersion (ORD) and circular dichroic (CD) properties are discussed under spectroscopic properties (Sec 1.5.3).

Table 1.3 The physical properties** of twenty coded α-amino acids

α-Amino acid	Molar mass (gmol⁻¹)	Decom. temp. (°C)	Solubility in water (g/100 g 25 °C)	Taste in H_2O (pH = 6.0)	$[\alpha]_D^{25}$ H_2O	$[\alpha]_D^{25}$ 5M HCl	Surface area† in Å²	van der waal's volume‡ in Å³
Alanine	89.10	297	16.5	sweet	+ 1.6	+ 13.0	115	67
Arginine	174.21	238	15.0§	bitter	+21.8	+ 48.1	225	167
Asparagine	132.12	236	3.1	bitter	−7.4	+ 37.8	160	148
Aspartic acid	133.11	270	0.5	bitter	+6.7	+ 33.8	150	67
Cysteine	121.16	178*	v.sol	sulphur -ous	−20.0	+ 7.9	135	86
Glutamic acid	147.14	249	0.84	tasty	+17.7	+ 46.8	190	109
Glutamine	146.15	185	3.6	sweet	+9.2	+ 46.5	180	114

Glycine	75.07	292	25	sweet	–	–	75	48
Histidine	155.16	277	7.59	bitter	–59.8	+18.3	195	118
Isoleucine	131.18	284	4.12	bitter	+16.3	+51.8	175	124
Leucine	131.18	337	2.3	bitter	–14.4	+21.0	170	124
Lysine	146.19	224	v.sol	flat	+19.7	+37.9	200	135
Methionine	149.22	283	3.5	tasty	–14.9	+34.6	185	124
Phenylalanine	165.20	284	2.97	bitter	–57.0	–7.4	210	135
Proline	115.14	222	162.3	sweet	–99.2	–69.5	145	90
Serine	105.10	228	5.0	sweet	–7.9	+15.9	115	73
Threonine	119.12	253	20.5	sweet	–33.9	–17.9	140	93
Tryptophan	204.23	282	1.14	bitter	–68.8	–5.7	255	163
Tyrosine	181.20	344	0.05	bitter	N.S.	–18.1	230	141
Valine	117.15	315	8.85	bitter	+6.6	+33.1	155	105

* as hydrochloride. § at 21°C

** Data adopted from 'The Protein Amino Acids' by P.M. Hardy in Chemistry and Biochemistry of the Amino Acids Ed. G.C. Barrett PP. 6-24, Chapman and Hall 1985.

†, ‡ From 'Proteins, Structure and Function' by David Whitford pp. 18-22, John Wiley & Sons, 2005.

1.5.2 Acid-Base Properties of Amino Acids

The amino acids are observed to have low solubilities and high melting points as compared to the compounds containing similar groups and having comparable molar mass as shown below.

Compound	Molar mass (g mol⁻¹)	Solubility		Melting point (°C)
		in water	in alcohol	
Lactic acid	90	very high	very high	~4
Alanine	89	moderate	insoluble	~297
3-Amino-2-butanol	89	very high	very high	~9

This difference can be explained in terms of the dipolar nature of the amino acid. This is due to the formation of an internal salt by movement of a proton from carboxylic group to amino group. This dipolar species is called a zwitterion and the amino acid is said to be in **zwitterionic** form.

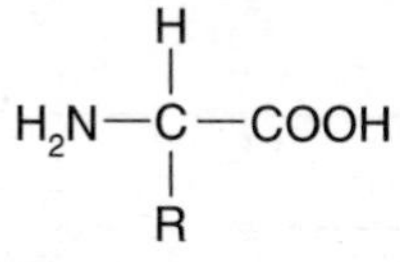

$$H_2N-\underset{\underset{R}{|}}{\overset{\overset{H}{|}}{C}}-COOH \qquad\qquad H_3\overset{+}{N}-\underset{\underset{R}{|}}{\overset{\overset{H}{|}}{C}}-COO^-$$

Unionised form 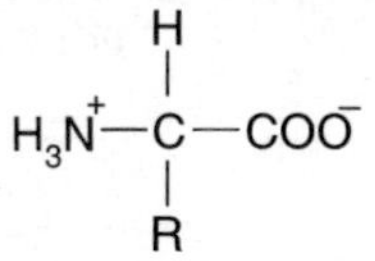   Zwitterionic form

In a crystal lattice the 'ionic' amino acids are held together by strong forces of intermolecular interactions. These contribute to the high melting and boiling points and solubility of amino acids. The presence of zwitterionic species is also responsible for the electrical conductivity of aqueous solutions of amino acids.

Since the amino acids contain acidic and basic functional groups and both of these groups are capable of ionisation in aqueous solutions, these show interesting acid-base properties. In water, amino acids can act both as acids as well as bases i.e., these are **amphoteric** in nature. The nature of predominant molecular species present in an aqueous solution of amino acid will depend on the pH of the solution. This aspect is very important in determining the reactivities of different amino acid side chains and their influence on the properties of the proteins.

The acid-base behaviour of amino acids depends on the nature of the side chain. The twenty coded amino acids may be put into two groups for this purpose. The first group includes those amino acids which do not have an ionisable group in the side chain and the second group of amino acids with ionisable group in the side chain.

1.5.2.1 Amino Acids with Non -Ionisable Side Chains

The acid-base behaviour of the amino acids with non-ionisable side chains can be understood by taking alanine as an example. The curve for titration of alanine with NaOH shows two inflection points–one around a pH of 2.3 and another around 9.7 (Fig. 1.11). These correspond to the pK_a values of the carboxyl group and the amino group respectively.

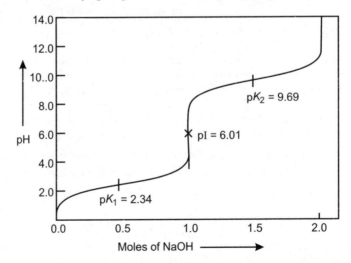

Fig. 1.11 *Titration curve for alanine–an amino acid with non-ionisable side chain.*

At a pH of less than 2, both the carboxyl and amino functional groups ·will be protonated, and the alanine molecule will have a net positive charge (species I). As we raise the pH of the solution, the carboxyl group gets deprotonated and we get a dipolar species II with no net charge. There would be an equilibrium between species I and species II, the relative amounts of the two being determined by the pH of the solution. The equilibrium can be represented as:

$$
\underset{\text{I}}{H_3\overset{+}{N}-\underset{\overset{|}{CH_3}}{\overset{\overset{H}{|}}{C}}-COOH} \; \rightleftharpoons \; \underset{\text{II}}{H_3\overset{+}{N}-\underset{\overset{|}{CH_3}}{\overset{\overset{H}{|}}{C}}-COO^-}
$$

This equilibrium would shift towards species II with further increase in the pH till a characteristic pH is reached, called the **isoelectric point (pI)** where species II would be predominant.

This behaviour is common to all amino acids other than the ones containing an acidic or a basic side chain. Thus isoelectric point, pI, may be defined as the pH of an aqueous solution of an amino acid at which its molecules on an average have no net charge. The pI for alanine is found to be about 6.0, which means that at a pH of 6.0, alanine would be primarily in the zwitterionic form.

The dipolar ion (species II) is also a potential acid as the $-NH_3^+$ group can donate a proton. Therefore, further rise in the pH would cause this ammonium group to deprotonate and generate an anionic form of the amino acid (species III). There would again be an equilibrium this time between species II and species III, the relative amounts of the two species again being determined by the pH of the solution. At a pH greater than 10, the anionic species III would be predominant.

$$
\underset{\text{II}}{H_3\overset{+}{N}-\underset{\overset{|}{CH_3}}{\overset{\overset{H}{|}}{C}}-COO^-} \; \rightleftharpoons \; \underset{\text{III}}{H_2N-\underset{\overset{|}{CH_3}}{\overset{\overset{H}{|}}{C}}-COO^-}
$$

The isoelectric point of alanine is related to the pK_a values of the carboxyl and the amino functional group and in fact it is the average of the two pK_a values.

$$pI = \frac{2.34 + 9.69}{2} = \frac{12.03}{2} = 6.015 \approx 6.0$$

The acid-base properties of the coded α-amino acids are compiled in Table 1.4.

Table 1.4   The acid-base properties of coded α-amino acids

α-Amino acid	pK_a; CO$_2$H (main chain) pK_1	pK_a; α-NH$_3^+$ pK_2	pK_a; side chain pK_R	Isoelectric pH
Alanine	2.34	9.69	–	6.01
Arginine	2.17	9.04	12.84	10.76
Asparagine	2.02	8.60	–	5.41
Aspartic acid	1.88	9.60	3.65	2.77
Cysteine	1.71	8.18	10.28	5.02
Glutamic acid	2.16	9.67	4.32	3.24
Glutamine	2.17	9.13	–	5.65
Glycine	2.34	9.60	–	5.97
Histidine	1.82	9.17	6.00	7.59
Isoleucine	2.36	9.68	–	6.02
Leucine	2.36	9.60	–	5.98
Lysine	2.18	9.12	10.53	9.82
Methionine	2.28	9.21	–	5.74
Phenylalanine	1.83	9.13	–	5.48
Proline	1.99	10.6	–	6.30
Serine	2.21	9.15	–	5.68
Threonine	2.71	9.62	–	6.16
Tryptophan	2.38	9.39	–	5.89
Tyrosine	2.20	9.11	10.07	5.66
Valine	2.32	9.62	–	5.96

In general, the isoelectric points of the amino acids can be computed by the following formula.

$$pI = \frac{1}{2}(pK_i + pK_j)$$

where, pK_i and pK_j refer to the pK_1 and pK_2 values i.e., the pK_a values for the main chain –COOH and the α-NH$_3^+$ group respectively for the neutral amino acids. However, for acidic amino acids these refer to pK_1 and pK_R

while for the basic amino acids these represent pK_R and pK_2 values respectively; pK_R being the pK_a value of the side chain functional group.

The presence of different charged species in an aqueous solution of α-amino acid can be shown experimentally by observing the movement of these species in an electric field, using the technique called **electrophoresis**. In such an experiment a gel containing a buffer solution of a given pH is set and a small amount of the amino acid is placed near the centre of the gel. Then an electric potential is applied at its ends. The schematic representation of the set-up of this experiment is given in Fig. 1.12.

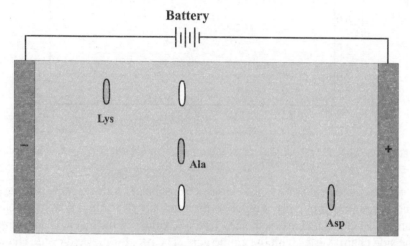

Fig. 1.12 *Schematic representation of electrophoresis of a mixture of three amino acids; lysine, alanine and aspartic acid in a gel buffered at a pH of 6.01. The blank spots indicate the position of lysine and aspartic acid before applying the electrical field. In case of alanine the position is same before and after applying the field.*

On applying the potential, different species present at the pH of the buffer move towards the electrodes depending on their net charge. In the experiment represented by Fig.1.12, three different amino acids are examined simultaneously at a pH of 6.01. It can be observed that the spot for aspartic acid moves towards anode implying that this amino acid has a net negative charge. Similarly the spot moving towards the cathode refers to an amino acid having a net positive charge i.e., lysine; while the amino acid spot that does not move indicates that the corresponding amino acid (alanine) is at its isoelectric point i.e., it is in the dipolar or zwitterionic form.

1.5.2.2 Amino Acids with Ionisable Side Chains

The acid-base behaviour of the amino acids with ionisable side chains is quite complicated. These show complex titration curves with three inflection points one each for the carboxyl group , the amino group and the side chain ionisable group. The titration curve for aspartic acid with ionisable carboxyl group in the side chain is shown in Fig. 1.13.

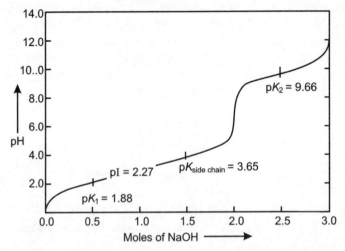

Fig. 1.13 *Titration curve for aspartic acid having an ionisable carboxyl group in the side chain.*

For amino acids with ionisable side chains the pK_a values for the carboxyl and the amino group are similar to the ones for the amino acids without ionisable side chains, while the pK_R (the pK_a values of the ionisable group in the side chain) varies with the amino acid. The pK_R for the acidic side chain of aspartic acid (3.65) is close to that of the pK_{COOH} while the basic side chain of lysine (pK_R = 10.53) shows the inflection point close to the amino group ionisation. The equilibria involved in the titration of aspartic acid and lysine can be represented as shown in Fig. 1.14.

(a)

(b)

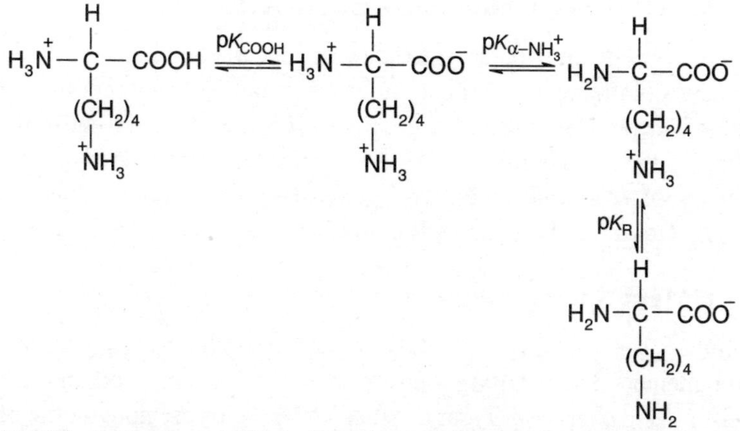

$$H_3\overset{+}{N}-\underset{\underset{\underset{NH_3}{\overset{+}{|}}}{\underset{(CH_2)_4}{|}}}{\overset{\overset{H}{|}}{C}}-COOH \underset{}{\overset{pK_{COOH}}{\rightleftharpoons}} H_3\overset{+}{N}-\underset{\underset{\underset{NH_3}{\overset{+}{|}}}{\underset{(CH_2)_4}{|}}}{\overset{\overset{H}{|}}{C}}-COO^- \underset{}{\overset{pK_{\alpha-NH_3^+}}{\rightleftharpoons}} H_2N-\underset{\underset{\underset{NH_3}{\overset{+}{|}}}{\underset{(CH_2)_4}{|}}}{\overset{\overset{H}{|}}{C}}-COO^-$$

$$pK_R \Updownarrow$$

$$H_2N-\underset{\underset{NH_2}{\underset{(CH_2)_4}{|}}}{\overset{\overset{H}{|}}{C}}-COO^-$$

Fig. 1.14 *The ionisation equilibria for (a) aspartic acid–an acidic amino acid and (b) lysine–a basic amino acid.*

The acid- base equilibria of the ionisable side chains of the amino acids are summarised in Table 1.5.

Table 1.5 The acid-base equilibria of the ionisable side chains of α-amino acids*

Amino acid residue (group)	Acid-base equilibrium of the side chain	Representative pK$_a$ value
Aspartic acid / Glutamic acid (side chain carboxyl)	$-COOH \rightleftharpoons -COO^- + H^+$ Carboxylic acid Carboxylate	4.4
Histidine (imidazole)	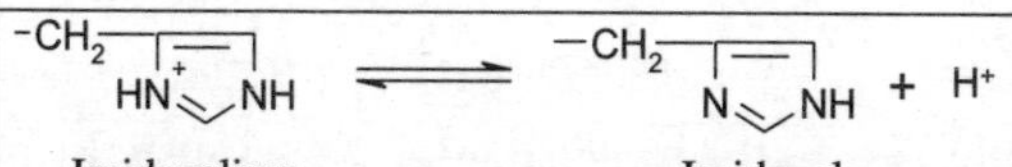 $-CH_2$—[imidazolium] $\rightleftharpoons$ $-CH_2$—[imidazole] $+ H^+$ Imidazolium Imidazole	6.5
Cysteine (thiol)	$-SH \rightleftharpoons -S^- + H^+$ Thiol Thiolate	8.5
Tyrosine (phenolic)	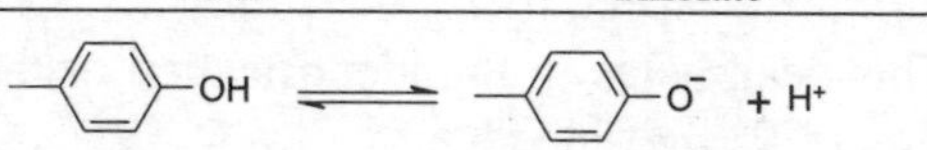 phenol-OH $\rightleftharpoons$ phenol-O$^-$ $+ H^+$ Phenol Phenolate	10.0
Lysine (side chain amino)	$-\overset{+}{N}H_3 \rightleftharpoons -NH_2 + H^+$ Ammonium Amine	10.0
Arginine (guanidinyl)	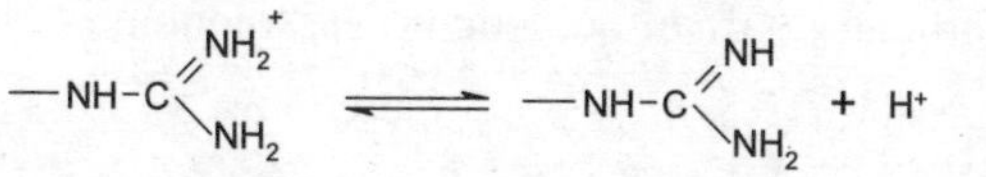 $-NH-C\overset{\overset{+}{N}H_2}{\underset{NH_2}{\diagdown}} \rightleftharpoons -NH-C\overset{NH}{\underset{NH_2}{\diagdown}} + H^+$ Guanidinium Guanidino	12.0

*   The main chain carboxyl group of amino acid has a pK$_a$ range of ~ 3.5-4.0 while for the α-amino group the range is ~ 8.0-9.0

1.5.3 Spectral Properties of α-Amino Acids

Spectroscopy, indisputably is the most versatile and indispensable tool in the hands of practising chemist. Since the spectroscopic properties of the peptides and proteins depend a great deal on that of the α-amino acids constituting them it is pertinent to have an understanding of the spectroscopic properties of α-amino acids. The general spectroscopic behaviour of α-amino acids is briefly described below.

1.5.3.1 Mass Spectrometry

In mass spectrometry the volatile molecules to be analysed are ionised and fragmented. The resulting ions are separated and analysed on the basis of their mass to charge (m/z) ratio. Since amino acids are not volatile, these are normally studied as their volatile derivatives like, esters. The fragmentation pattern of these derivatives is quite simple to analyse. There are two fragmentation pathways, both involving the cleavage of C–C bond α to the amino group. These are referred to as α-cleavages. The stability of the resulting immonium ions determines the preferred pathway.

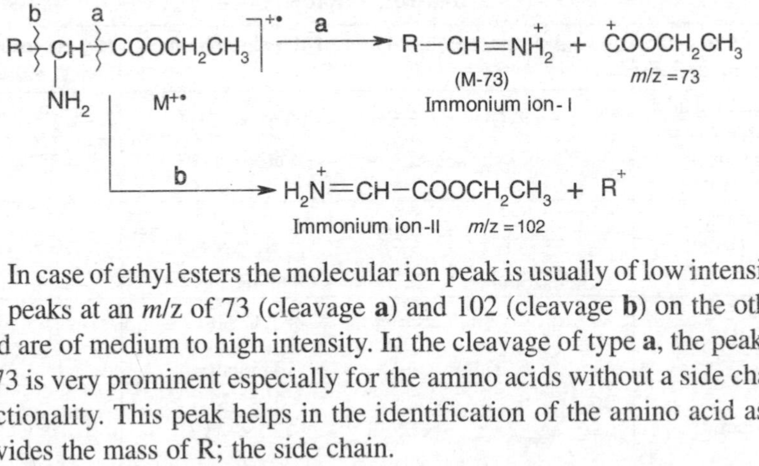

In case of ethyl esters the molecular ion peak is usually of low intensity. The peaks at an m/z of 73 (cleavage **a**) and 102 (cleavage **b**) on the other hand are of medium to high intensity. In the cleavage of type **a**, the peak at M-73 is very prominent especially for the amino acids without a side chain functionality. This peak helps in the identification of the amino acid as it provides the mass of R; the side chain.

When the side chain contains two or more aliphatic carbon atoms the M-73 ion ejects an alkene as a consequence of a hydrogen migration from farther carbon atoms and generates a strong peak at m/z of 30. Thus a peak at m/z of 30 is indicative of valine, leucine or isoleucine.

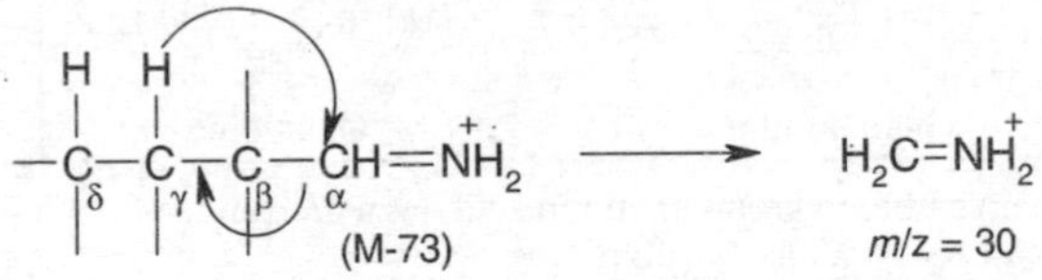

In case of leucine , there is a possibility of **Mc Lafferty rearrangement** also to give a peak at m/z of 44.

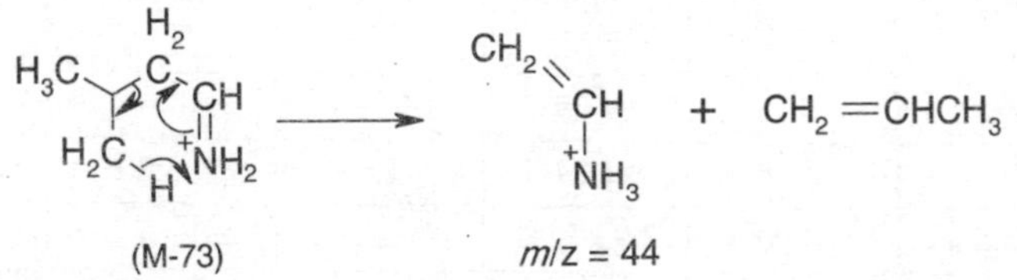

In type **b** cleavage, the M-102 peak (due to R^+) is suggestive of the mass of the side chain. Further, the m/z 102 peak can also undergo a Mc Lafferty rearrangement to give a peak at m/z of 74.

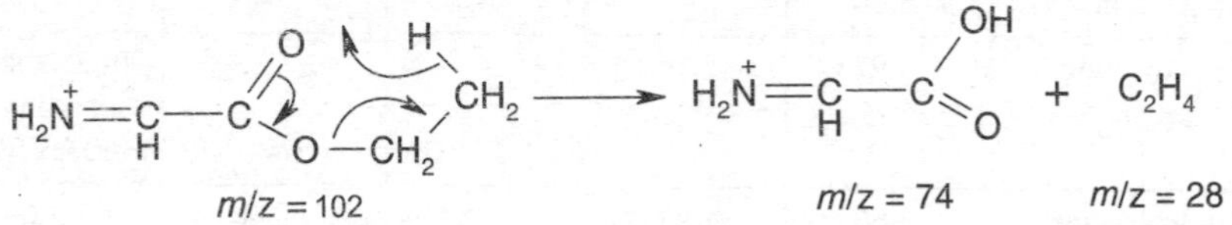

In case the side chain of the amino acid contains a functional group then the fragmentation pattern may be similar to the one discussed above with additional fragments. Sometimes the fragmentation pattern may be competitive with the general pattern discussed above.

1.5.3.2 Nuclear Magnetic Resonance (NMR) Spectroscopy

The NMR spectroscopy is extensively being exploited in the study of amino acids, peptides and proteins. The objectives vary from routine analysis to the conformational aspects and structure determination. However, the structural details of the peptides revealed by NMR are second only to the X-ray crystallographic analysis.

The chemical shift positions of the protons (^1H) and carbon (^{13}C) nuclei are found to depend significantly on the nature and state of ionisation of different groups of a given amino acid i.e., whether the amino acid is in the cationic, zwitterionic or the anionic form. In case of peptides and proteins the chemical shift positions depend on the nature and state of ionisation of functional groups of the neighboring residues besides depending on solvent and temperature etc. The chemical shift positions of α, β and other protons of the side chains of different amino acids are compiled in Table 1.6 .

Table 1.6 The PMR chemical shifts of α-amino acids[*, **]

α-Amino acid	α-Hydrogen (ppm)	β-Hydrogen (ppm)	γ-Hydrogen (ppm)	δ-Hydrogen (ppm)	Other protons (ppm)
1. Alanine	3.78	1.49	–	–	–
2. Arginine	3.18	1.87	1.6	3.20	
3. Asparagine	4.00	2.92, 2.87	–	–	–
4. Aspartic acid	4.08	3.01	–	–	–
5. Cysteine	3.82	2.93	–	–	–
6. Glutamic acid	3.75	2.12	2.49	–	–
7. Glutamine	3.77	2.16	2.43	–	–
8. Glycine	3.56	–	–	–	–
9. Histidine	3.97	3.16	–	–	Imidazole protons 7.03, 7.72
10. Isoleucine	3.66	1.97	1.35	1.00	ε = 1.00
11. Leucine	3.70	1.69, 1.73	1.71	0.96	ε = 0.96
12. Lysine	3.37	1.69	1.43	1.69	ε =2.95
13. Methionine	3.80	2.60	2.12	2.12	–
14. Phenylalanine	3.97	3.20	–	–	Aromatic protons 7.36
15. Proline	4.08	2.06	2.06	–	–
16. Serine	3.94	3.94	–	–	–
17. Threonine	3.58	4.23	1.32	–	–
18. Tryptophan	4.04	3.38	–	–	Indole protons 7.31, 7.22, 7.28, 7.55
19. Tyrosine	3.93	3.05, 3.17	–	–	Aromatic protons 7.19, 6.88
20. Valine	3.62	2.29	1.06	1.06	–

*   Data adopted from Nuclear Magnetic Spectra of Amino Acids by G.C.Barrett and
     J.S.Davies in Chemistry and Biochemistry of Amino Acids Ed. G.C.Barrett, pp 537-
     539, Chapman and Hall, 1985.
** The values are for zwitterionic form in D_2O and are relative to DSS (2,2-dimethyl-2-
     silapentane-5-sulphonate) as internal standard.

1.5.3.3 UV Spectroscopy

The common functional groups (amino, carboxyl and amide) present in the amino acids and peptides are transparent in the UV region, therefore amino acids with aliphatic side chains have no absorption above 220 nm in the ultraviolet range. However, the chromophores present in the side chains of a few coded aromatic amino acids viz., Tyr, Trp and Phe do absorb in the range of ultraviolet spectrum. The λ_{max} positions and the corresponding molar absorptivity values for these aromatic amino acids are given in Table 1.7.

Table 1.7 λ_{max} and molar absorptivity values for aromatic amino acids at pH = 6.0

Amino acid	λ_{max} (in nm)	Molar absorptivity, ε (in $cm^2\ mol^{-1}$)
Tyrosine	193	48000
	225	8000
	272	1200
Tryptophan	218	35000
	281	5500
	188	60000
Phenylalanine	208	8000
	260	150
Histidine	211	5900
Cysteine	250	300

This allows the application of UV spectroscopy in the quantification of peptides and proteins containing these residues. The pH dependent spectra of proteins can provide information about number and nature of the tyrosyl residues present in the protein. These chromophores provide 'optical handles' to study the conformational aspects of peptides and proteins by the circular dichroic measurements also. Further, the tyrosyl and tryptophanyl side chains show fluorescence spectra in the UV range. For example, tyrosine containing peptides and proteins show an emission maxima around 310-315 nm on being excited at 275 nm.

1.5.3.4 IR Spectroscopy

The absence of characteristic NH stretching (3300-3500 cm^{-1}) and the carbonyl absorption for –COOH group (1700-1730 cm^{-1}) in the IR spectra of amino acids is indicative of the zwitterionic nature of amino acids. These characteristic frequencies instead are replaced by absorptions around $3070 \, cm^{-1}$ and $1560\text{-}1600 \, cm^{-1}$ that are characteristic of the NH_3^+ and COO^- respectively. The NH_3^+ group has additional bands in the $1500\text{-}1600 \, cm^{-1}$ region. In addition to these, most of the amino acids show a medium absorption around $1300 \, cm^{-1}$ and weak absorptions in $2080\text{-}2140 \, cm^{-1}$ and $2530\text{-}2760 \, cm^{-1}$ regions. When these amino acids combine to give peptides a number of additional bands characteristic of the peptide bond and of the conformation adopted by the peptide are also observed.

1.5.3.4 Circular Dichroism (CD) ,

In water, the C_α alkylated amino acids show a positive **cotton effect** with a peak around 200-204 nm at neutral pH. This band called **band 1** shifts to higher wavelengths in acidic and basic pH. In addition, these C_α alkylated amino acids show a negative cotton effect band called **band 2** around 245-250 nm in acidic pH range. The aromatic amino acids (except histidine) and the sulphur containing amino acids, on the other hand, show a maxima above 250 nm in their CD spectra. The CD spectra, like the UV spectra, are sensitive to the state of ionisation of different functional groups in the amino acid.

1.6 CHEMICAL REACTIONS OF AMINO ACIDS

Since amino acids have both an amino as well as a carboxyl group, these show the reactions of the two groups individually as well as in combination. These reactions are outlined below.

1.6.1 Reactions due to Amino Group

1. **With alkyl halides:** Amino acids form alkyl derivatives on treatment with alkyl halides. In presence of excess alkyl halide, these form internal quaternary alkyl ammonium salts. These salts are zwitterionic in nature and are called **betaines.** The term betaine is derived from a naturally occurring compound bearing the same name. Chemically it is N, N, N-trimethylglycine which can be prepared from glycine in presence of excess methyl iodide.

$$H_3\overset{+}{N}-CH_2-COO^- \xrightarrow{\quad 3 \ CH_3I \quad} (CH_3)_3\overset{+}{N}-CH_2-COO^-$$

Glycine   N,N,N-Trimethylglycine (betaine)

Similarly the imino acid, proline containing cyclised side chain reacts in the following manner to give a dimethyl derivative.

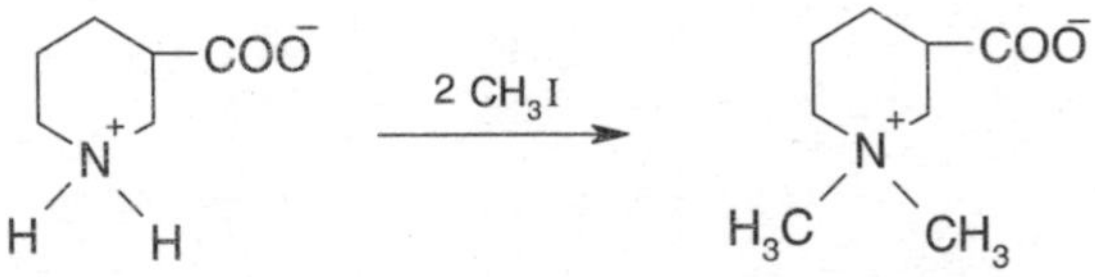

Proline   N,N-Dimethylproline

2. **With nitrous acid:** In the reaction of primary aliphatic amines with nitrous acid the primary amino group is lost as nitrogen through the formation of corresponding diazonium group. The N_2 gas is evolved in quantitative yields. Amino acids with primary amino groups also undergo reaction with nitrous acid in a similar manner and produce corresponding hydroxy acid.

$$R-CH-COOH + HNO_2 \longrightarrow R-CH-COOH + N_2$$
$$\quad\ \ | \qquad\qquad\qquad\qquad\qquad\ \ |$$
$$\quad\ \ NH_2 \qquad\qquad\qquad\qquad\quad OH$$

Amino acid   α-Hydroxyacid

This reaction forms the basis for **Von Slyke determination** of amino nitrogen which in turn gives an estimate of the amount of amino acid.

3. **With acid anhydride:** Amino acids react with acid anhydrides and acid chlorides to produce acyl amino acids. The reaction is carried out under sufficiently alkaline conditions so that a substantial concentration of the free amino group is present. However it is necessary to acidify the aqueous solution to obtain the acidic product.

$$H_2NCH_2COOH \xrightarrow{(CH_3CO)_2O} CH_3CONHCH_2COOH$$

Glycine   N-Acetylglycine

$$H_2NCHCOOH \xrightarrow[\substack{\text{OH , H}_2\text{O, 2h, 4°C} \\ \text{(ii) HCl}}]{\text{(i) } C_6H_5COCl} C_6H_5CONHCHCOOH$$
$$\quad\ \ | \qquad\qquad\qquad\qquad\qquad\qquad\qquad |$$
$$\quad\ CH(CH_3)_2 \qquad\qquad\qquad\qquad\qquad CH(CH_3)_2$$

Valine   N-Benzoylvaline

When α-amino acids are treated with trifluoroacetic anhydride the initial amide rapidly cyclises and the hydrolysis of the cyclised product gives α-keto acid.

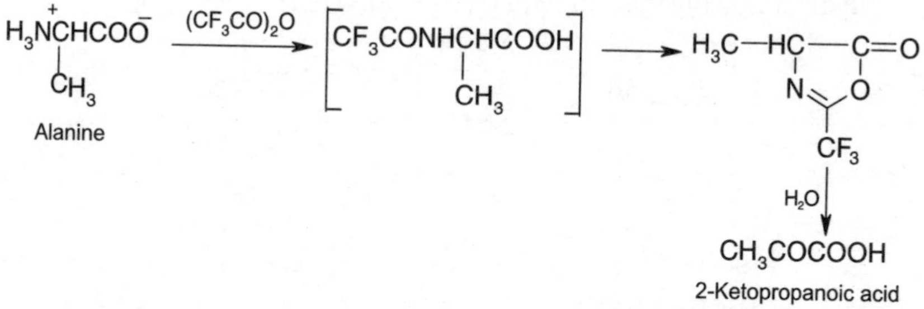

$$\underset{\substack{| \\ CH_3 \\ \text{Alanine}}}{H_3\overset{+}{N}CHCOO^-} \xrightarrow{(CF_3CO)_2O} \left[\underset{\substack{| \\ CH_3}}{CF_3CONHCHCOOH} \right] \longrightarrow \underset{\substack{| \\ N \quad O \\ \backslash\;/ \\ C \\ | \\ CF_3}}{H_3C-HC-C=O}$$

$$\Big\downarrow H_2O$$

$$CH_3COCOOH$$

2-Ketopropanoic acid

The acyl derivatives have an added advantage that these may be treated with $SOCl_2$ to give acid chlorides, an important step in peptide synthesis.

$$CH_3CONHCH_2COOH + SOCl_2 \longrightarrow CH_3CONHCH_2COCl + HCl + SO_2$$

N-Acetylglycine

Formation of acid chlorides by direct treatment of amino acid with PCl_5 or $SOCl_2$ is otherwise difficult because of the presence of amino group.

4. **With sulphonyl azide:** The amino group of the amino acids reacts with sulphonyl azide under mild conditions and gets converted into azido group.

$$\underset{\text{Leucine}}{H_2NCH(CH_2CHMe_2)COO^-} + \underset{\text{Sulphonyl azide}}{CF_3SO_2N_3} \xrightarrow[\text{pH=9,12h}]{H_2O \,/\, CH_2Cl_2}$$

$$\underset{\text{Azido-leucine}}{N_3CH(CH_2CHMe_2)COO^-}$$

The reaction is significant because it proceeds without loss or change of optical activity.

5. **With formaldehyde:** Amino acids undergo nucleophilic addition with formaldehyde to form a mixture of products.

$$\underset{\text{Glycine}}{\overset{+}{N}H_3-CH_2-COO^-} \xrightarrow{CH_2O} \underset{\text{Methyleneglycine}}{H_2C=N-CH_2COOH} + \underset{\text{Dimethylolglycine}}{(CH_2OH)_2N-CH_2COOH}$$

In the derivatives so obtained the amino group of the amino acid is blocked and the carboxyl group can be titrated with a standard alkali to determine the amount of the amino acid. Such a titration is called **Sorenson's formol titration**. A direct titration of the amino acid is not possible due to interference from the free amino group.

6. **With nitrosyl halides:** On reaction with nitrosyl chloride and bromide,

the amino acids give the respective halo acids.

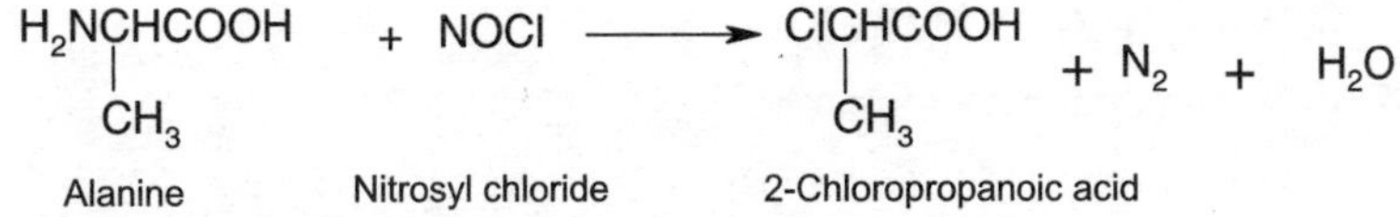

$$\underset{\underset{CH_3}{|}}{H_2NCHCOOH} \; + \; NOCl \longrightarrow \underset{\underset{CH_3}{|}}{ClCHCOOH} \; + \; N_2 \; + \; H_2O$$

Alanine   Nitrosyl chloride   2-Chloropropanoic acid

7. **With hydroiodic acid :** On reacting with hydroiodic acid at 200°C, amino acids lose their amino group as a molecule of ammonia to give an acid. For example,

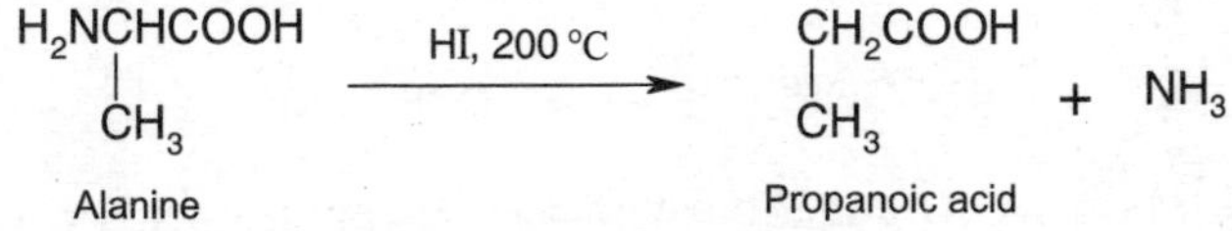

$$\underset{\underset{CH_3}{|}}{H_2NCHCOOH} \xrightarrow{\text{HI, 200 °C}} \underset{\underset{CH_3}{|}}{CH_2COOH} \; + \; NH_3$$

Alanine   Propanoic acid

1.6.2 Reactions due to Carboxyl Group

1. **Esterification :** The carboxyl group of amino acids can be easily esterified by treating a suspension of amino acid in an appropriate alcohol with anhydrous hydrogen chloride. The ester product formed can give back the amino acid on treatment with cold dilute alkali. The reaction of methyl alcohol with alanine is shown below.

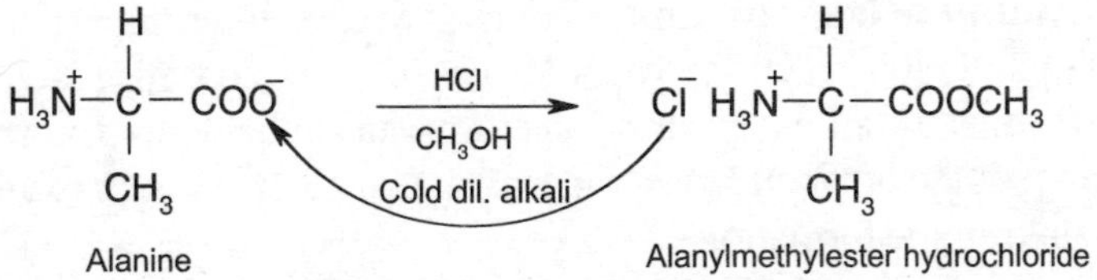

$$\underset{\underset{CH_3}{|}}{\overset{\overset{H}{|}}{H_3\overset{+}{N}-C-COO^-}} \underset{\underset{\text{Cold dil. alkali}}{\underset{CH_3OH}{\xleftarrow{\hspace{1cm}}}}}{\xrightarrow{\text{HCl}}} Cl^- \; \underset{\underset{CH_3}{|}}{\overset{\overset{H}{|}}{H_3\overset{+}{N}-C-COOCH_3}}$$

Alanine   Alanylmethylester hydrochloride

When benzene sulphonic acid is used as a catalyst , benzyl esters are obtained. Water produced in the reaction is removed by azeotropic distillation, avoiding the use of large excess of benzyl alcohol. For example, glycine forms the following benzyl ester on reacting with benzyl alcohol in presence of benzene sulphonic acid.

$$H_3\overset{+}{N}-CH_2-COO^- + \langle \!\!\!\bigcirc\!\!\! \rangle -CH_2OH \xrightarrow{C_6H_5SO_3H}$$

Glycine

$$C_6H_5SO_3^- \; [\, H_3\overset{+}{N}-CH_2COO-CH_2-\langle \!\!\!\bigcirc\!\!\! \rangle$$

Glycinebenzylester benzenesulphonate

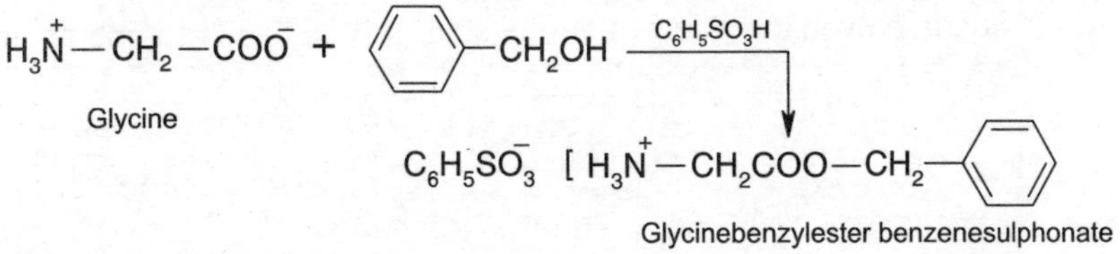

The product formed above may be reacted with H_2 in presence of palladium to give the amino acid salt from which the free amino acid can be obtained.

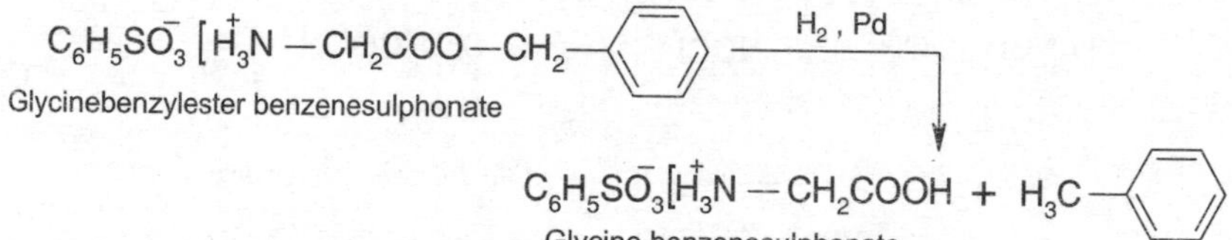
$C_6H_5SO_3^-$ [$H_3\overset{+}{N}$ — CH_2COO—CH_2—
H_2, Pd
$C_6H_5SO_3^-$[$H_3\overset{+}{N}$ — CH_2COOH + H_3C—
Glycinebenzylester benzenesulphonate

Glycine benzenesulphonate

2. **Decarboxylation :** Amino acids lose a molecule of carbon dioxide from the carboxyl group on heating with barium peroxide and yield amines.

$$H_2N-CH(R)COOH \xrightarrow{\Delta,\ BaO_2} RCH_2-NH_2 + CO_2$$

3. **Reduction :** On reduction with $LiAlH_4$ amino acids form amino alcohols with the retention of optical activity.

$$H_2N-CH(R)COOH \xrightarrow{LiAlH_4} H_2N-CH(R)CH_2OH$$

1.6.3 Reactions due to Both Amino and Carboxyl Groups

1. **Pyrolysis :** On heating amino acids different products are obtained depending on the position of the amino group.

(i) **α-Amino acids :** α-Amino acids give 2,5-diketopiperazines on heating. In this reaction two molecules of α-amino acids react in such a way that the amino group of one forms an amide with the carboxylic group of the other and vice-versa as shown below. It is an example of double amide formation.

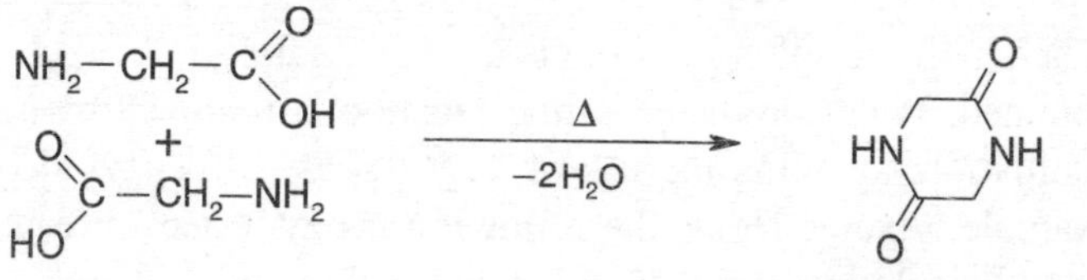
NH_2—CH_2—C
O
OH
+
O
C—CH_2—NH_2
HO
Δ
−2H_2O
O
HN
NH
O
Glycine

2,5-Diketopiperazine

On heating, even the esters of amino acids give 2,5-diketopiperazine.

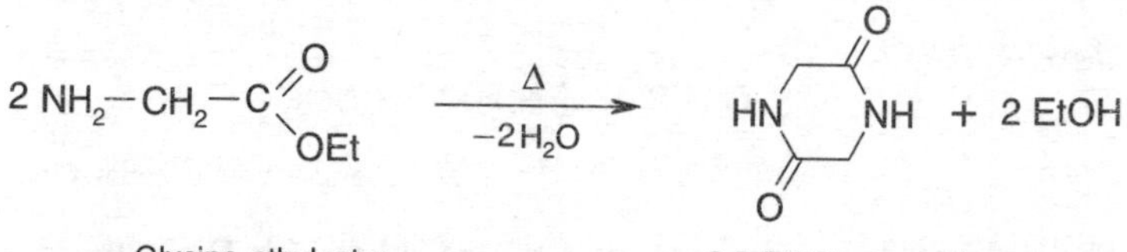
2 NH_2—CH_2—C
O
OEt
Δ
−2H_2O
O
HN
NH
+ 2 EtOH
O
Glycine ethylester

2,5-Diketopiperazine

(ii) **β-Amino acids :** On heating, β-amino acids eliminate a molecule of ammonia to give α, β-unsaturated acids.

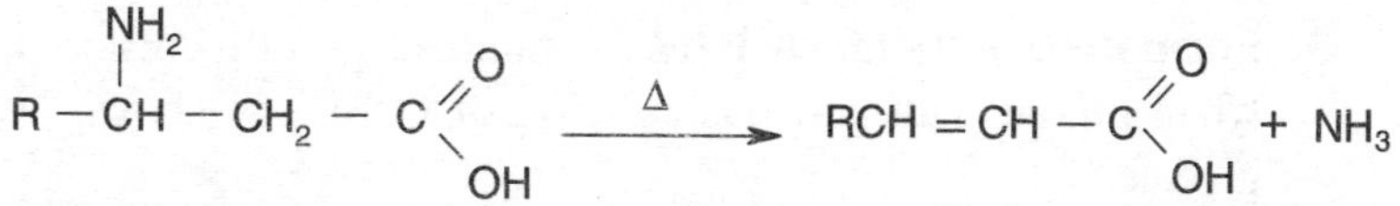

$$\underset{\beta\text{-Amino acid}}{R-\overset{\overset{\displaystyle NH_2}{|}}{CH}-CH_2-C\overset{\displaystyle O}{\underset{\displaystyle OH}{\diagup}}} \xrightarrow{\Delta} RCH=CH-C\overset{\displaystyle O}{\underset{\displaystyle OH}{\diagup}} + NH_3$$

(iii) γ-Amino acids and δ-Amino acids : γ- and δ- amino acids on heating form cyclic amides called lactams. The lactams formed show a special type of tautomerism called **lactam-lactim tautomerism**.

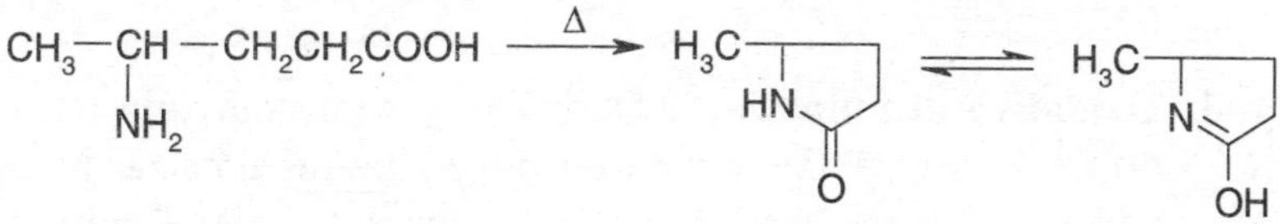

$$\underset{\text{4-Aminopentanoic acid}}{CH_3-\overset{\overset{\displaystyle }{|}}{\underset{\underset{\displaystyle NH_2}{|}}{CH}}-CH_2CH_2COOH} \xrightarrow{\Delta} \underset{\text{Lactam}}{H_3C} \rightleftharpoons \underset{\text{Lactim}}{H_3C}$$

Linear polymeric amides are formed when amino and carboxylic groups are far from each other along the chain.

$$H_2N(CH_2)_n COOH + H_2N(CH_2)_n COOH \longrightarrow$$

$$H_2N-\left[(CH_2)_n-CONH\right]_m-COOH$$

2. **Formation of hydantoin :** With isocyanates, α-amino acids form carbamides which form hydantoin on warming with HCl.

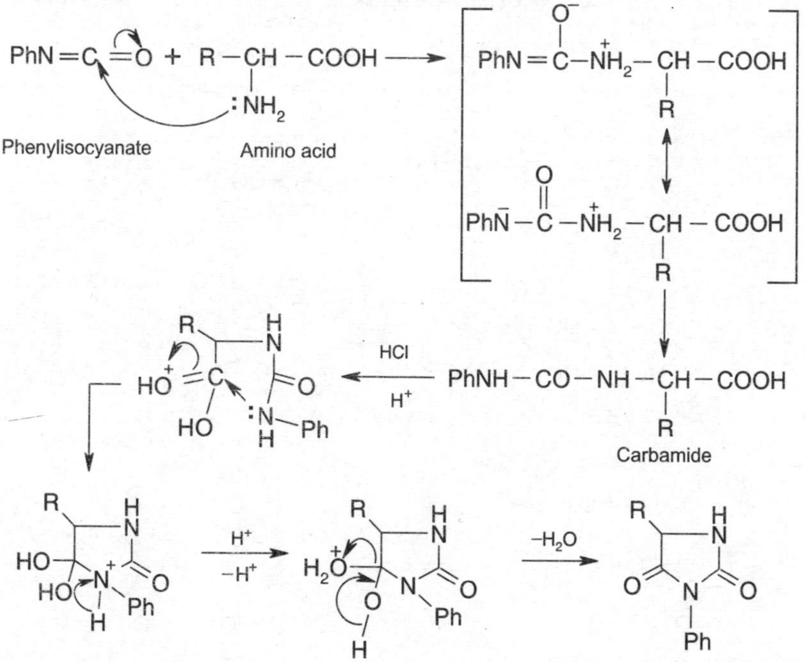

$$PhN=C=O + \underset{\text{Amino acid}}{R-\overset{\overset{\displaystyle }{|}}{\underset{\underset{\displaystyle :NH_2}{}}{CH}}-COOH} \longrightarrow \left[PhN=C-\overset{+}{NH_2}-\overset{\overset{\displaystyle }{|}}{\underset{\underset{\displaystyle R}{}}{CH}}-COOH \right.$$

Phenylisocyanate

$$\left. Ph\bar{N}-\overset{\overset{\displaystyle O}{||}}{C}-\overset{+}{NH_2}-\overset{\overset{\displaystyle }{|}}{\underset{\underset{\displaystyle R}{}}{CH}}-COOH \right]$$

$$\xrightarrow{} \underset{\text{Carbamide}}{PhNH-CO-NH-\overset{\overset{\displaystyle }{|}}{\underset{\underset{\displaystyle R}{}}{CH}}-COOH}$$

$$\xleftarrow[\text{H}^+]{\text{HCl}}$$

$$\xrightarrow[-H^+]{H^+}$$

$$\xrightarrow{-H_2O}$$

Hydantoin

3. **Formation of metal chelates :** Amino acids form chelate structures when mixed with aqueous solution of $CuSO_4$ as shown below for glycine.

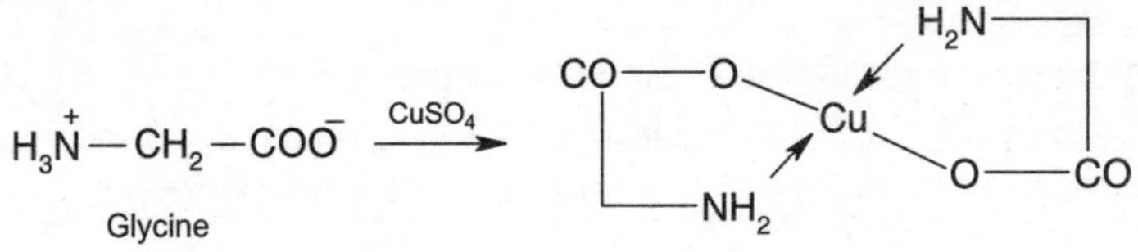

$$H_3\overset{+}{N}-CH_2-\overset{-}{COO} \xrightarrow{CuSO_4}$$

Glycine

bis- Glycinato Cu(II)

4. **Oxidative deamination :** On reacting with ninhydrin (triketohy-drindene hydrate) the á-amino acids gets oxidised into an imino acid which undergoes deamination, releasing ammonia. The reaction is called **oxidative deamination** and is also the path of biodegradation of amino acids. The likely reaction sequence is as given below.

The liberated ammonia reacts with hydrindantin to give the coloured product. This reaction forms the basis of detection of amino acids in their chromatographic determination wherein ninhydrin acts as a spray or detection agent.

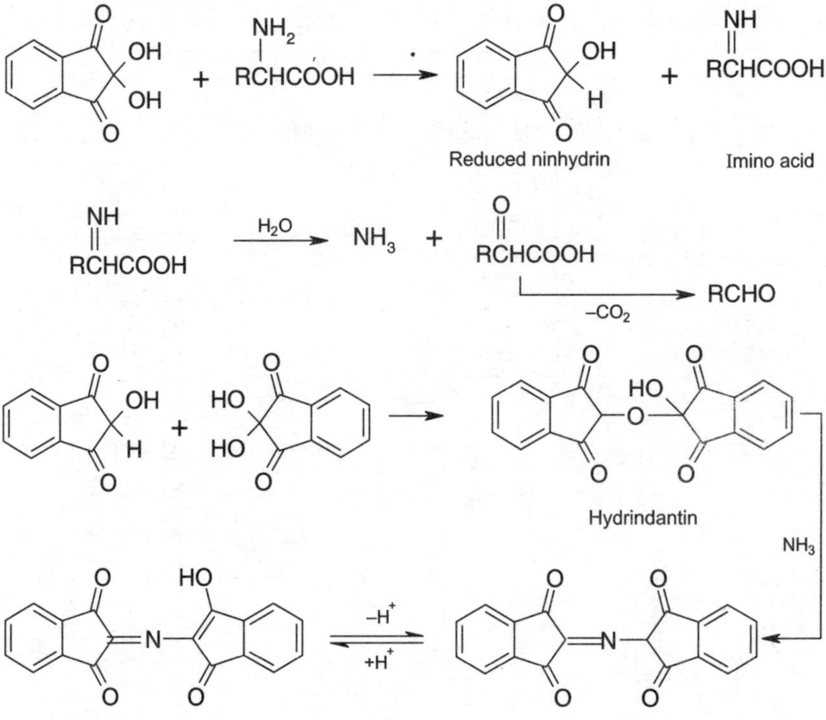

Reduced ninhydrin   Imino acid

Hydrindantin

Violet coloured product

1.7 INDUSTRIAL PREPARATION OF α-AMINO ACIDS

The manufacturing methods of amino acids can be categorised as:

- Extraction from hydrolysates of animal or plant proteins
- Chemical synthesis
- Fermentation method

The two methods viz., extraction from hydrolysates of animal and plant proteins and fermentation have been discussed below. The chemical synthesis being elaborate has been discussed separately under section 1.8.

Extraction from Hydrolysates of Animal or Plant Proteins

Traditionally, amino acids were obtained by separating the mixture of amino acids obtained on hydrolysis of plant and animal proteins. In this method, proteins are extracted from plant or animal sources and are purified by dialysis. The free protein (devoid of ionic impurities) is then dried and subjected to acidic or alkaline hydrolysis or to enzymatic degradation with proteolytic enzymes. The separation of the amino acids from protein hydrolysate is then achieved with the help of separation procedures like, electrophoresis, ion-exchange or paper chromatography etc.

Though this method provides moderate quantities of coded and post-transitionally modified α-amino acids it is quite tedious and somewhat expensive. Besides, we cannot get all the amino acids by this method as a number of amino acids are either destroyed or modified during the process of hydrolysis (Sec. 3.4.1.1). More so with the availability of economically viable synthetic methods this traditional method is losing importance.

Fermentation Method

The fermentation method is used for the commercial production of a number of amino acids. The manufacturing process by fermentation generally comprises of three stages viz.,

- Fermentation
- Crude isolation
- Purification

In the first step of fermentation process, a microorganism is grown in a suitable fermentation medium. In the next step, most impurities contained in the fermentation broth are removed and in the third step, final purification is performed to obtain the desired product. In a typical production of L-amino acid by direct fermentation, the bacterium is grown in a suitable fermentation medium consisting of a carbon source (e.g., glucose, molasses, alkanes, glycerol, ethanol etc.), ammonia as a nitrogen source, and a small amount of minerals and vitamins as growth factors in a clean and sterile fermentation tank. The pH, temperature and dissolved oxygen content are the control factors in the fermentation process and are to be maintained. Once the growth of the bacteria reaches a certain level, the amino acid starts accumulating in the medium. After the fermentation process is over the broth is centrifuged or filtered through a membrane filter to separate bacterial cells and proteins. Crude crystals of the desired amino acid are then obtained by crystallisation of the supernatant or filtrate so obtained. Sometimes it is difficult to obtain the crystals in this way. In such cases, first the impurities are removed by means of an ion-exchange resin or activated carbon and then the filtrate is subjected to crystallisation.

1.8 CHEMICAL SYNTHESIS OF α-AMINO ACIDS

A constant supply of optically pure α- amino acids and their analogues is required as these (or the peptides obtained from them) perform important biological functions and also have pharmaceutical importance. For example. α-methyl-3′, 4′-dihydroxyphenylalanine is used in the treatment of Parkinson's disease.

The coded amino acids can be obtained from protein hydrolysates by using suitable separation methods as discussed above. However, this method of obtaining amino acids is quite tedious and expensive. Further, non-coded amino acids cannot be obtained from the protein hydrolysates. Therefore, industrially viable methods for synthesising α-amino acids are required. A wide spectrum of methods is available for synthesising amino acids. Some of the commonly used methods of synthesising α-amino acids are given below.

1. **Amination of α-halo acids :** This is the most general method involving displacement reactions of α-halo acids. It involves amination of a α-halo acid with excess of ammonia (**direct ammonolysis**). However, not all amino acids can be synthesised by this method.

$$\underset{\substack{\text{X} \\ \text{α-Halo acid}}}{\text{RCHCOOH}} \xrightarrow{\text{NH}_3\text{(excess)}} \underset{\substack{\overset{+}{\text{NH}_3} \\ \text{α-Amino acid}}}{\text{RCHCOO}^-} + \text{NH}_4\text{X}$$

The reaction can also be visualised as alkylation of ammonia with alkyl α-halo carboxylic acids. In this S_N2 reaction, ammonia acts as a nucleophile. Since ammonia can be alkylated to di- and trialkyl stage, the reaction gives a complex mixture of products. Therefore, this method of preparing α-amino acids is usually not preferred. However, by using excess of ammonia one can get more of monoalkylation product. For example,

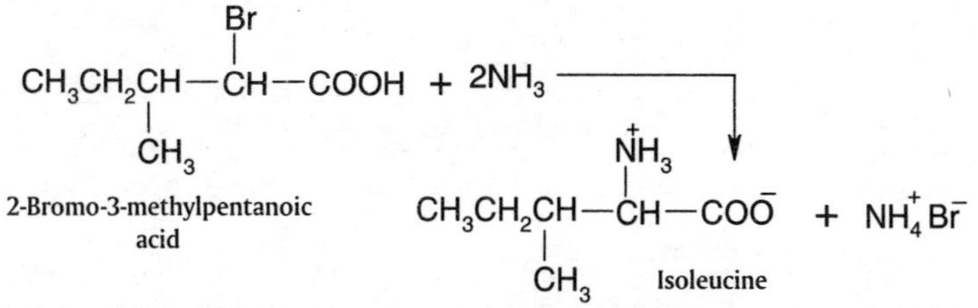

$$CH_3CH_2CH(CH_3)-\overset{Br}{\underset{}{CH}}-COOH + 2NH_3 \longrightarrow$$

2-Bromo-3-methylpentanoic acid

$$CH_3CH_2CH(CH_3)-\overset{\overset{+}{NH_3}}{\underset{}{CH}}-CO\bar{O} + NH_4^+Br^-$$

Isoleucine

α-Halo acids can be conveniently prepared by the **Hell-Volhard-Zelinsky (HVZ) reaction.** Here, carboxylic acids react with chlorine or bromine in presence of phosphorus or phosphorus halide to yield the α-halo acid. Halogenations occur specifically at the α-carbon because the reaction probably proceeds via enolisation of the acyl halide. The function of phosphorus halide is to convert a little of the acid into acid halide.

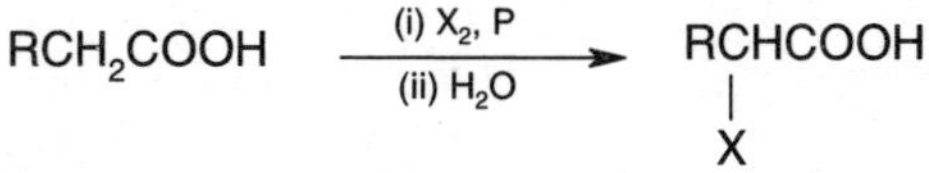

$$RCH_2COOH \xrightarrow[\text{(ii) } H_2O]{\text{(i) } X_2, P} RCHCOOH$$
$$| \atop X$$

For example,

$$CH_3CH_2C(=O)OH \xrightarrow{Br_2, P} \left[CH_3CHC(=O)Br \atop | \atop Br \right] \xrightarrow{H_2O} CH_3CHC(=O)OH \atop | \atop Br$$

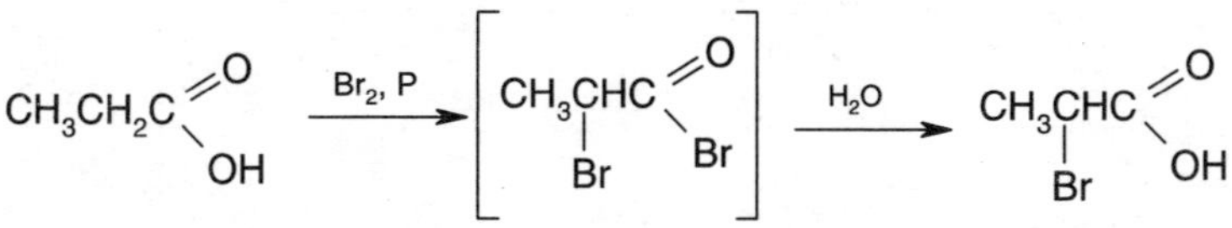

If more than one molar equivalent of the halogen is taken then depending upon the availability of α-hydrogen atoms, di or trihalo acids are obtained.

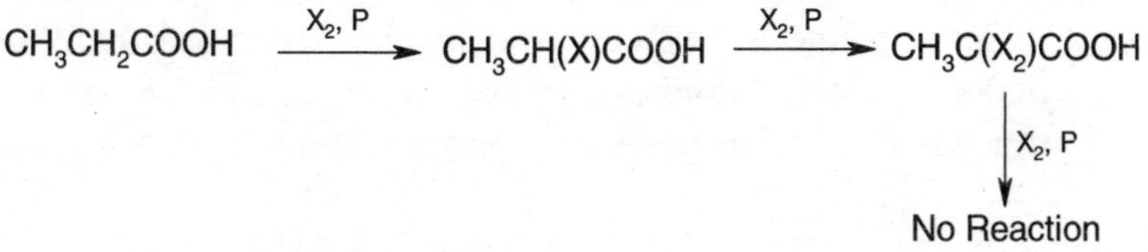

$$CH_3CH_2COOH \xrightarrow{X_2, P} CH_3CH(X)COOH \xrightarrow{X_2, P} CH_3C(X_2)COOH$$
$$\downarrow X_2, P$$
$$\text{No Reaction}$$

Alternatively, α-haloacids can also be prepared by using modified malonic ester synthesis. In this method an ester is taken as the reactant instead of an acid. The reaction is given as:

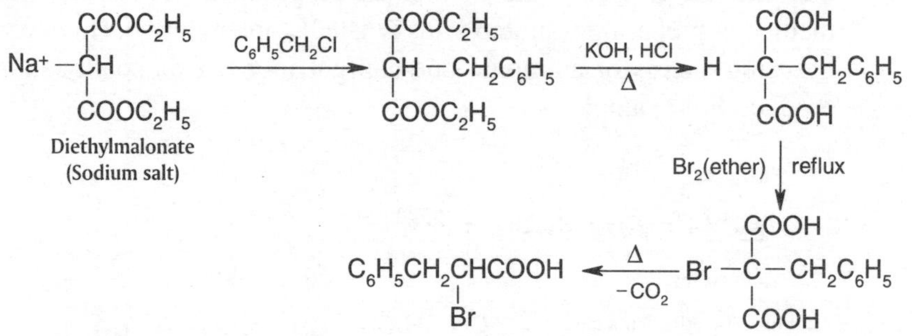

Amino acids like glycine, alanine, valine, leucine, isoleucine and aspartic acid, serine and threonine can be sysnthesised using direct ammonolysis method. The drawback of this method is that the yields are poor. The method given below is used to obtain better yields.

2. **Using potassium pthalimide:** This method is a modification of Gabriel pthalimide synthesis of amines and gives better yields. It is generally used for the preparation of glycine and leucine. In this method α-haloesters are used instead of α-haloacids. The reaction between α-haloester and potassium phthalimide is carried out in the following steps to obtain α-amino acids.

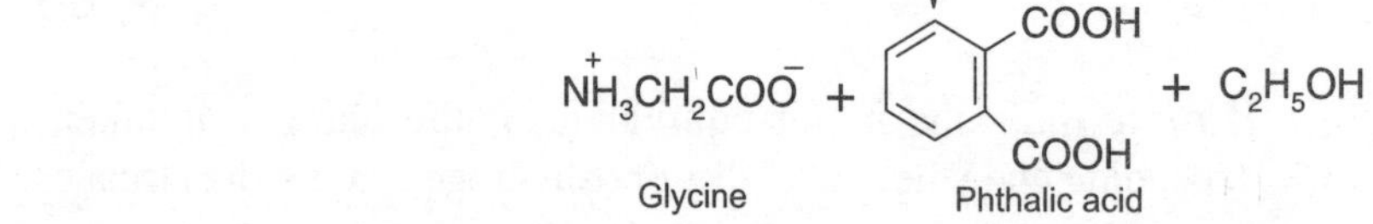

This method can be modified to carry out the synthesis of other amino acids like valine, isoleucine and methionine. It is called as **phthalimido malonic ester method** and makes use of diethyl-α-bromomalonate instead of ethylchloroacetate to prepare an imidomalonic ester.

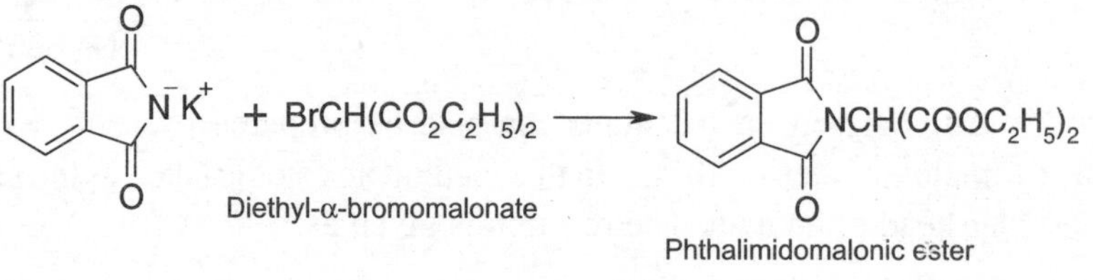

The amino malonate derivative (phthalimidomalonic ester) so obtained is then alkylated. For this, the ester is treated with sodium ethoxide in ethanol to generate an anion which then acts as a nucleophile in the S_N2 reaction with alkyl halide to give the alkyl derivative of malonic ester.

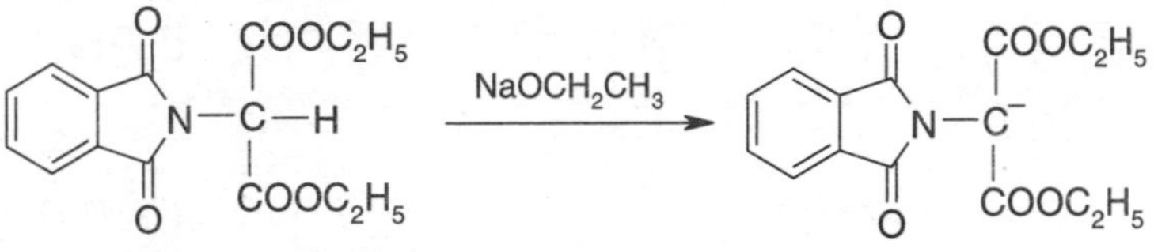

Phthalimidomalonic ester

Chloroethylmethyl sulphide

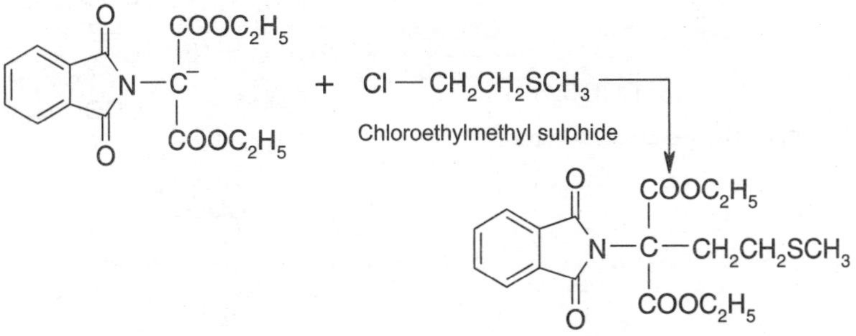

Saponification of the alkyl derivative followed by acidification gives substituted amino malonic acid which decarboxylates to generate an α-amino acid.

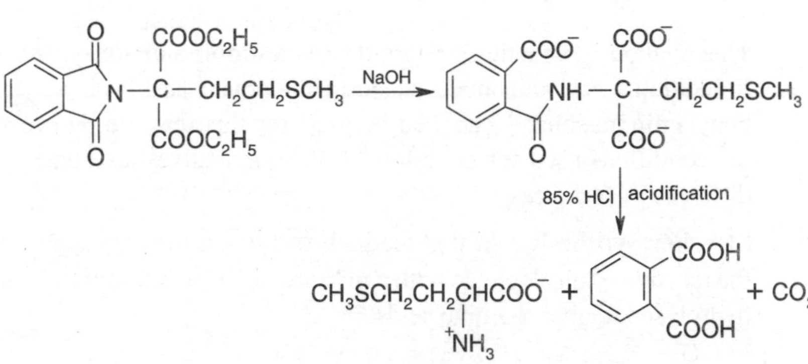

DL-Methionine

In place of phthalimido derivatives, other amino derivatives (acyl amino) of malonic ester may also be used for preparing α-amino acids. For example, diethyl-α-acetamidomalonate is treated with sodium ethoxide followed by an alkyl /aryl halide to give α-amino acid. Diethylacetamidomalonate can be easily obtained from diethyl malonate as follows.

$$CH_2(COOC_2H_5) \xrightarrow[\text{acetic acid}]{NaNO_2/} HON=C(COOC_2H_5) \xrightarrow[\text{(ii) acetic anhydride}]{\text{(i) } H_2/Pd(C)}$$

Diethylmalonate

$$CH_3-\overset{\overset{\displaystyle O}{\|}}{C}-NH-\underset{\underset{\displaystyle COOC_2H_5}{|}}{\overset{\overset{\displaystyle COOC_2H_5}{|}}{C}}-H$$

Diethylacetamidomalonate

$$CH_3-\overset{\overset{\displaystyle O}{\|}}{C}-NH-\underset{\underset{\displaystyle COOC_2H_5}{|}}{\overset{\overset{\displaystyle COOC_2H_5}{|}}{C}}-H \xrightarrow[C_2H_5OH]{NaOCH_2CH_3} CH_3-\overset{\overset{\displaystyle O}{\|}}{C}-NH-\underset{\underset{\displaystyle COOC_2H_5}{|}}{\overset{\overset{\displaystyle COOC_2H_5}{|}}{C^-}}$$

Diethyl -α-acetamidomalonate   Enolate ion

$$C_6H_5CH_2Cl$$

$$CH_3-\overset{\overset{\displaystyle O}{\|}}{C}-NH-\underset{\underset{\displaystyle COOC_2H_5}{|}}{\overset{\overset{\displaystyle COOC_2H_5}{|}}{C}}-CH_2-\bigcirc + Cl^-$$

Heat | H_2O, HBr

$$2\ EtOH + CH_3COOH + CO_2 + \left[H_3\overset{+}{N}-\underset{\underset{\displaystyle COOH}{|}}{CH}-CH_2-\bigcirc\right] Br^-$$

Phenylalanine hydrobromide

This method is sometimes referred to as **amidomalonate synthesis**. In addition to acylaminomalonate, formamidomalonates, benzyloxycarbonylaminomalonates can also be used for this reaction. However, the conditions used for the removal of N-derivatives are somewhat different in these cases.

3. **Strecker synthesis :** In this method, α-aminonitriles (prepared by the reaction of an aldehyde with a mixture of HCN and ammonia) are hydrolysed to give α-amino acids.

$$R-\overset{\overset{\displaystyle O}{\|}}{C}-H + NH_3 + HCN \longrightarrow \underset{\underset{\displaystyle NH_2}{|}}{RCHCN} \xrightarrow[H_2O]{H_3O^+,\ Heat} \underset{\underset{\displaystyle \overset{+}{N}H_3}{|}}{RCH\overline{COO}}$$

α-Aminonitrile   α-Amino acid

For example, acetaldehyde reacts with a mixture of HCN and ammonia to give 2-aminopropanenitrile, which on hydrolysis gives alanine.

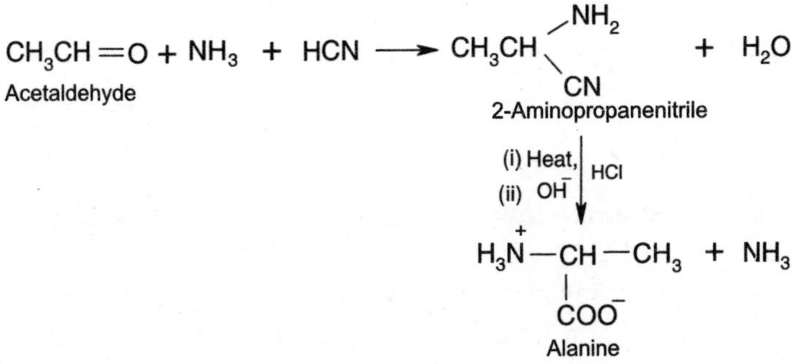

$$CH_3CH=O + NH_3 + HCN \longrightarrow CH_3CH \overset{NH_2}{\underset{CN}{\diagdown}} + H_2O$$

Acetaldehyde

2-Aminopropanenitrile

(i) Heat, HCl
(ii) OH⁻

$$H_3\overset{+}{N}-\underset{\underset{COO^-}{|}}{CH}-CH_3 + NH_3$$

Alanine

Glycine, leucine, isoleucine, valine, serine etc. can also be prepared using Strecker synthesis.

As far as the mechanism of the formation of aminonitrile is concerned, there are two possibilities. In one of the mechanisms the first step involves formation of imines from the aldehyde and ammonia which is followed by addition of HCN to yield the aminonitrile.

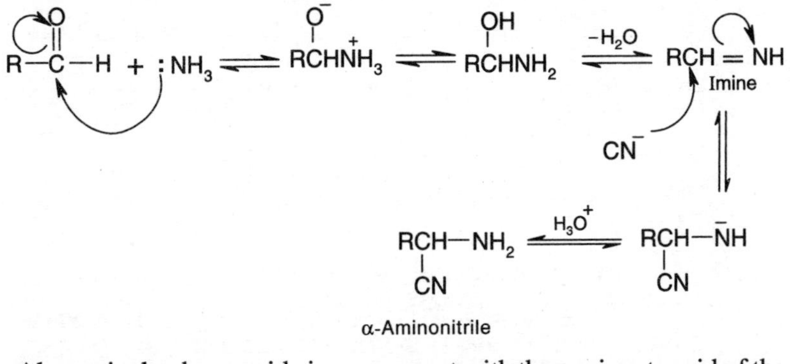

α-Aminonitrile

Alternatively, the cyanide ion may react with the conjugate acid of the imine to give the nitrile.

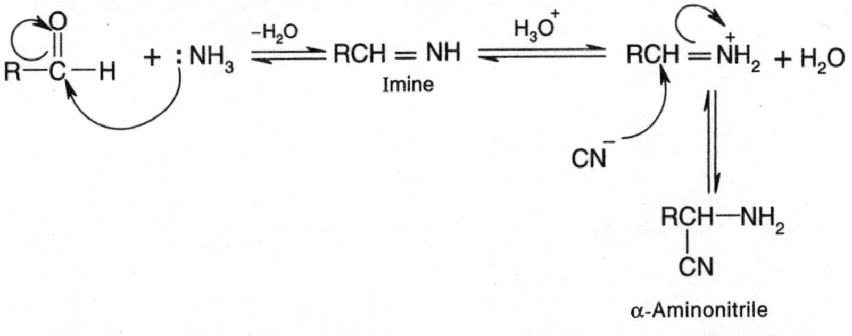

α-Aminonitrile

In the second mechanism the cyanohydrin may be produced first followed by a nucleophilic substitution.

$$R-\overset{\displaystyle O}{\underset{}{C}}-H + C\bar{N} \rightleftharpoons R-\underset{\underset{\displaystyle CN}{|}}{CH} \rightleftharpoons R-\underset{\underset{\displaystyle CN}{|}}{\overset{\overset{\displaystyle OH}{|}}{CH}} \xrightarrow[-H_2O]{:NH_3} R-\underset{\underset{\displaystyle CN}{|}}{\overset{\overset{\displaystyle NH_2}{|}}{CH}}$$

Cyanohydrin

4. **Reductive amination :** α-Amino acids can also be prepared by the reduction of aldehydes and ketones in the presence of ammonia. The method is called reductive amination. The reduction is carried out using a catalyst or sodium cyanohydridoborate, $NaBH_3CN$. For example, leucine can be prepared in the following manner starting from ethylisovalerate and ethyloxalate.

$$CH_3CHCH_2COOEt + \underset{\underset{\displaystyle COOEt}{|}}{COOEt} \xrightarrow{NaOEt} (CH_3)_2CHCH\overset{\displaystyle COOEt}{\underset{\displaystyle \underset{\displaystyle O}{\|}}{C\ COOEt}}$$

Ethylisovalerate   Ethyloxalate

$$(CH_3)_2\ CHCH_2CH-CO\bar{O} \xleftarrow[NH_3,\ H_2]{Pd/\Delta} \underset{\underset{\displaystyle NH_3^+}{|}}{} (CH_3)_2CHCH_2COCOOEt$$

with $\overset{-CO_2}{\underset{10\%\ H_2SO_4}{}} \Big| \overset{-C_2H_5OH}{\underset{boil}{}}$

Other α-amino acids prepared by this method are alanine, glutamic acid etc.

5. **Biosynthetic amination of α-keto acids :** α- keto acids may also be reductively aminated using ammonia and a reducing agent. This pathway is quite similar to the biosynthetic pathway.

$$CH_3-\overset{\displaystyle O}{\overset{\|}{C}}-COOH \xrightarrow[NaBH_4]{NH_3} CH_3-\underset{\underset{\displaystyle NH_2}{|}}{CH}-COOH$$

α-Ketopropanoic acid   Alanine

The biosynthetic pathway being,

$$CH_3-\overset{\displaystyle O}{\overset{\|}{C}}-COOH + NH_3 \xrightarrow{Pyridoxal-5'-phosphate} CH_3-\underset{\underset{\displaystyle NH_2}{|}}{CH}-COOH$$

6. **Reduction of azlactones :** Azlactones, prepared by condensation of aromatic aldehydes with N-acyl derivatives of glycine in presence of acetic anhydride and sodium acetate, can be reduced by phosphorus

and HI or other reducing agents to produce amino acids also increasing the chain by two, in the process.

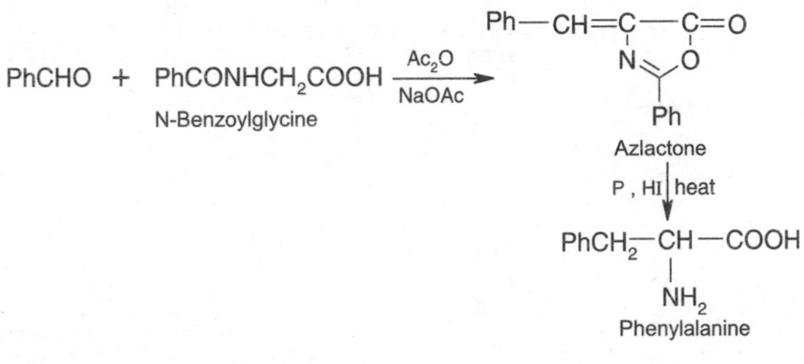

$$Ph-CH=\underset{\substack{N\\ \diagdown}}{C}-\underset{\substack{\diagup\\ O}}{C}=O$$

PhCHO + PhCONHCH$_2$COOH $\xrightarrow[\text{NaOAc}]{\text{Ac}_2\text{O}}$

N-Benzoylglycine

Ph

Azlactone

P , HI | heat

$$PhCH_2-\underset{\substack{|\\ NH_2}}{CH}-COOH$$

Phenylalanine

Alternatively, the azlactone is warmed with 1% NaOH to open the ring. The resulting product is reduced with sodium amalgam followed by hydrolysis to give the α-amino acid. This method is good for the preparation of aromatic amino acids like, phenylalanine, tyrosine etc. for example,

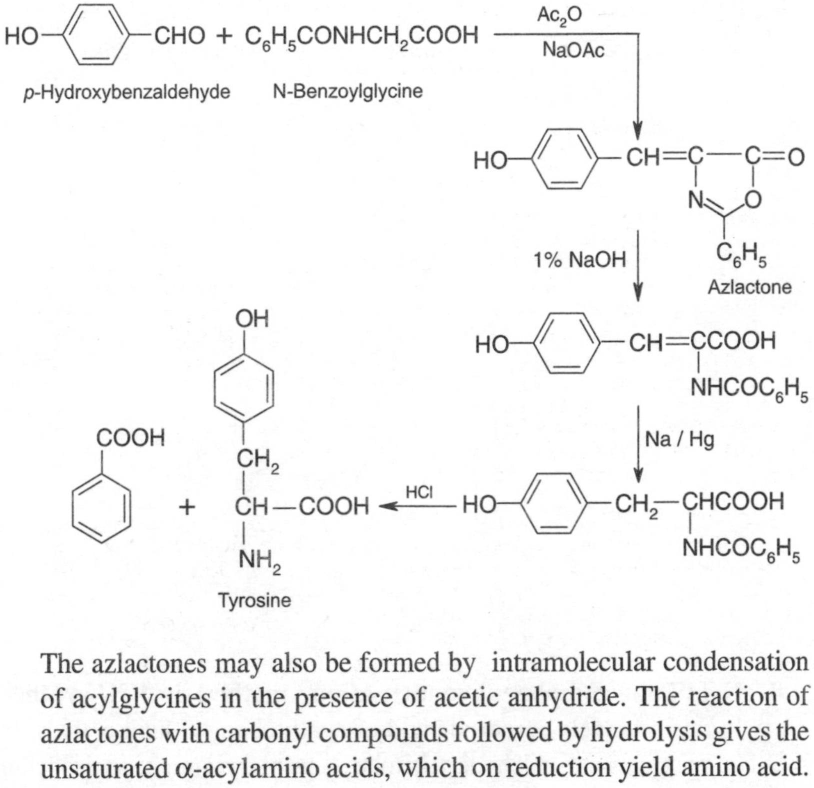

HO—⟨ ⟩—CHO + C$_6$H$_5$CONHCH$_2$COOH $\xrightarrow[\text{NaOAc}]{\text{Ac}_2\text{O}}$

p-Hydroxybenzaldehyde   N-Benzoylglycine

$$HO-\left\langle\ \right\rangle-CH=\underset{\substack{N\\ \diagdown}}{C}-\underset{\substack{\diagup\\ O}}{C}=O$$

1% NaOH | C$_6$H$_5$

Azlactone

$$HO-\left\langle\ \right\rangle-CH=\underset{\substack{|\\ NHCOC_6H_5}}{C}COOH$$

Na / Hg

$$HO-\left\langle\ \right\rangle-CH_2-\underset{\substack{|\\ NHCOC_6H_5}}{CH}COOH$$

$$\xleftarrow{\text{HCl}}$$

COOH
⟨ ⟩

+

OH
⟨ ⟩
CH$_2$
CH—COOH
NH$_2$

Tyrosine

The azlactones may also be formed by intramolecular condensation of acylglycines in the presence of acetic anhydride. The reaction of azlactones with carbonyl compounds followed by hydrolysis gives the unsaturated α-acylamino acids, which on reduction yield amino acid.

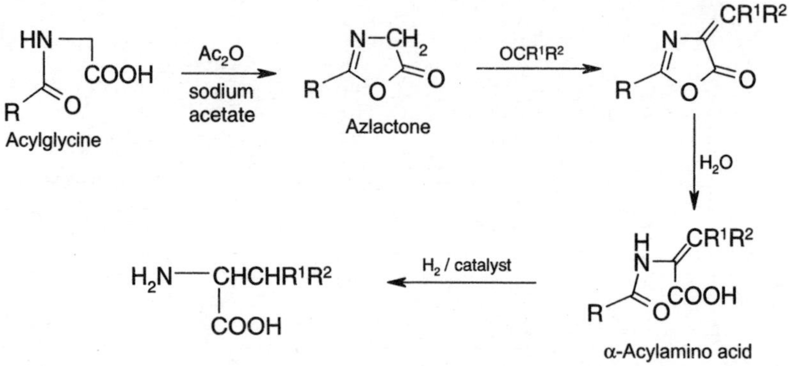

7. **By rearrangement of N-haloiminoesters :** N-haloiminoesters undergo Neber type of rearrangement in the presence of a base to first form an azirine intermediate and then α-amino orthoesters or α-amino esters. Both of these hydrolyse to give α-amino acids. The reactions can be given as under.

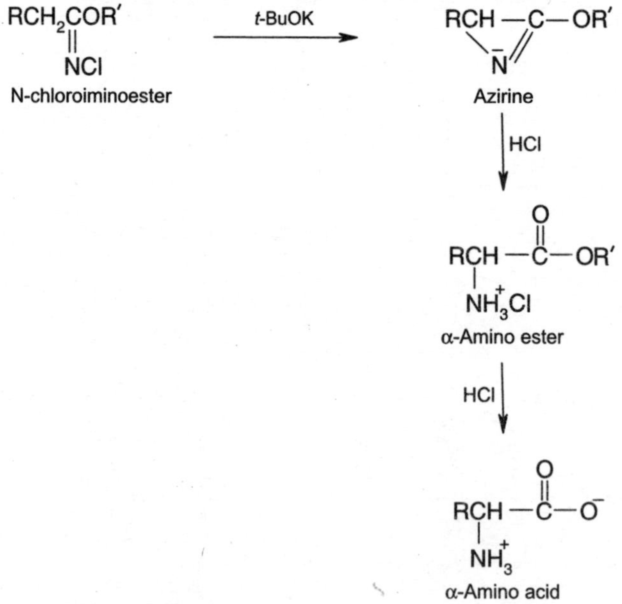

8. **Ugi "Four component condensation" method or 4CC synthesis:** In this relatively recent method of synthesising α-amino acids, four components–an aldehyde / ketone, a primary amine, an isocyanide

and a carboxylic acid 'condense' to form N^{α}-acylamino acid N-alkyl amide. The groups attached to the amino and the carboxyl groups are then removed to get the α-amino acid.

$$R^1NC + R^2R^3CO + R^4NH_2 + R^5COOH \longrightarrow R^5CONR^4CR^3R^2CONHR^1$$

N^{α}-acylaminoacid N^{α}-alkylamide

$$R^5CONR^4CR^3R^2CONHR^1 \xrightarrow[\text{(ii) } H_2 / Pd]{\text{(i) HBr}} H_2N-CH-COOH$$
$$\underset{R^3}{|}$$

For example, alanine can be synthesised as follows:

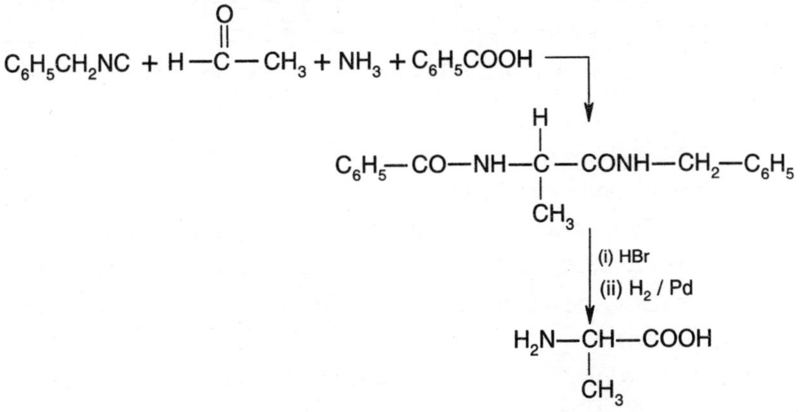

$$C_6H_5CH_2NC + H\overset{\overset{O}{\|}}{-C}-CH_3 + NH_3 + C_6H_5COOH \longrightarrow$$

$$C_6H_5-CO-NH-\overset{\overset{H}{|}}{\underset{\underset{CH_3}{|}}{C}}-CONH-CH_2-C_6H_5$$

$$\xrightarrow[\text{(ii) } H_2 / Pd]{\text{(i) HBr}}$$

$$H_2N-CH-COOH$$
$$\underset{CH_3}{|}$$
Alanine

This method though proves to be quite advantageous in preparing certain amino acids, it has not been able to replace the conventional methods of synthesising α-amino acids.

9. **Bucherer-Berg's synthesis :** In this method, α-amino acids can be prepared by the hydrolysis of a hydantoin obtained from the reaction of an aldehyde, with KCN and $(NH_4)_2CO_3$ as given below.

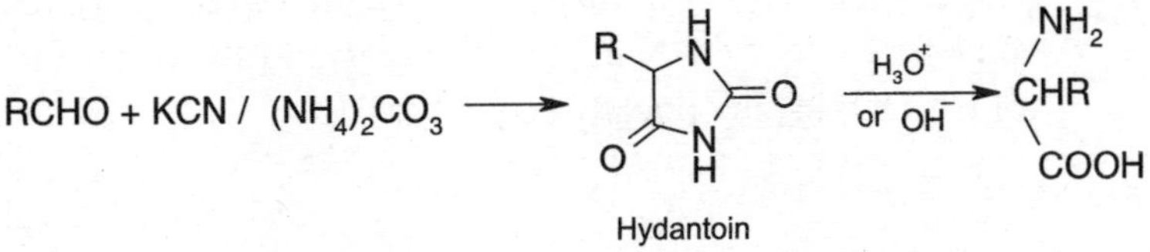

$$RCHO + KCN / (NH_4)_2CO_3 \longrightarrow \text{Hydantoin} \xrightarrow[\text{or OH}^-]{H_3O^+} \overset{NH_2}{\underset{COOH}{CHR}}$$

Hydantoin

10. **Miller-Urey method :** In this method atmospheric components like nitrogen, methane and water combine in presence of a high energy source to give a mixture of amino acids, most of them being α-amino acids.

$$N_2 + CH_4 + H_2O \xrightarrow{\text{Energy source}} \text{Mixture of amino acids}$$

(Mainly α-amino acids)

However, the yields are quite low and it is not possible to prepare a desired amino acid preferentially.

1.8.1 Enantiomeric Resolution of α-Amino Acids

All the amino acids, except glycine, synthesised by the above methods are obtained as racemic mixtures. It is necessary to resolve these mixtures in order to get a pure D or L enantiomer. The methods used for resolution are discussed below.

(a) **Amine salt formation :** In this method the racemic mixture is mixed with an enantiomer (either R or S) of a naturally occurring optically active amine like, strychnine or brucine. This leads to the formation of a diastereomeric pair of salts, which can be separated on the basis of differences in their physical properties like, solubility (crystallisation), melting point etc. The separated diastereomers are identified spectroscopically using circular dichroism etc. These separated salts on acidification precipitate respective amino acids. The separation with R-amine can be represented as follows.

Racemic mixture
Amino acids (R,S) $+ (R)-\text{Amine} \longrightarrow$ Diastereomeric salt mixture
(R,R) and (S,R)

Separation

$(R)-$ and $(S)-$Amino acids $\xleftarrow[\text{Hydrolysis}]{H_3O^+}$ $(R,R)-$Salt $+ (S,R)-$Salt

In place of direct salt formation, the racemic amino acids may first be converted into their N-acyl derivatives and then separated by salt formation with an optically active base.

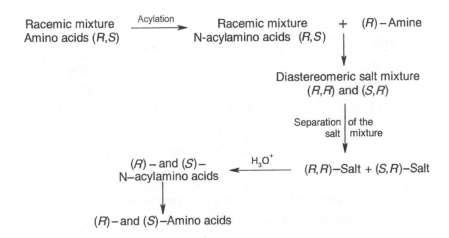

(b) Ester formation : In a similar method, the racemic mixture of amino acids may be converted into diastereomeric esters which are then separated by crystallisation and the amino acids recovered by hydrolysis.

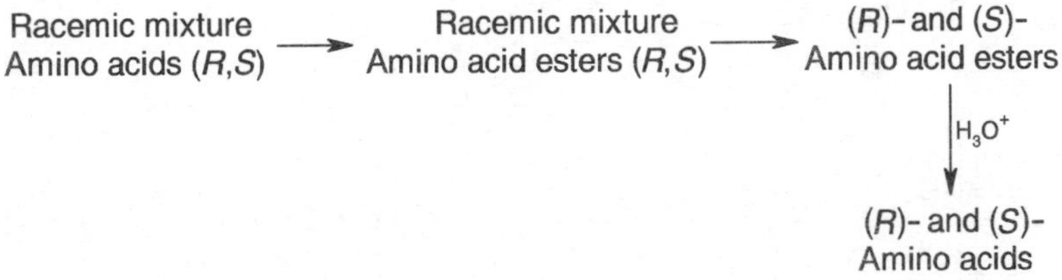

Racemic mixture   ⟶   Racemic mixture   ⟶   (R)- and (S)-
Amino acids (R,S)   Amino acid esters (R,S)   Amino acid esters

$\downarrow H_3O^+$

(R)- and (S)-
Amino acids

(c) Enzymic resolution of amino acids : The biological catalysts i.e., the enzymes can be used to resolve a racemic mixture of amino acids. Certain enzymes (called *deacylases*) obtained from living organisms can selectively catalyse the hydrolysis of one of the enantiomeric N-acylamino acid. For example, the enzyme obtained from hog kidney cleaves the acyl group from the L-enantiomer while the D-enantiomer remains unaffected. Thus to resolve a racemic mixture of amino acids it is converted into an N-acyl derivative and then the mixture is hydrolysed with the help of the deacylase enzyme.

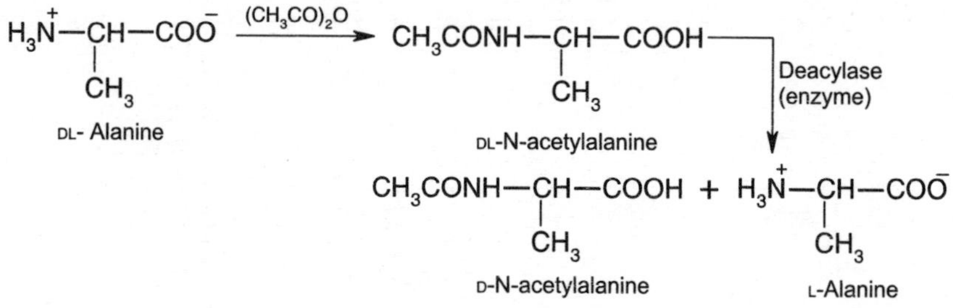
H3N+—CH—COO−
CH3
DL- Alanine
(CH3CO)2O
CH3CONH—CH—COOH
CH3
DL-N-acetylalanine
Deacylase
(enzyme)
CH3CONH—CH—COOH + H3N+—CH—COO−
CH3
CH3
D-N-acetylalanine
L-Alanine

The liberated amino acid is then precipitated from ethanol while the N-acyl derivative remains in solution.

(d) Chiral ligand exchange(LE) liquid chromatography (LC): It is a common chromatographic method employing chiral stationary phases for the separation of enantiomers. The chiral discrimination is attributed to the exchange of one stationary selector ligand in the bidentate complex on the stationary phase by an anylate ligand forming a ternary mixed complexes. The complex formation may be represented as shown in Fig.1.15.

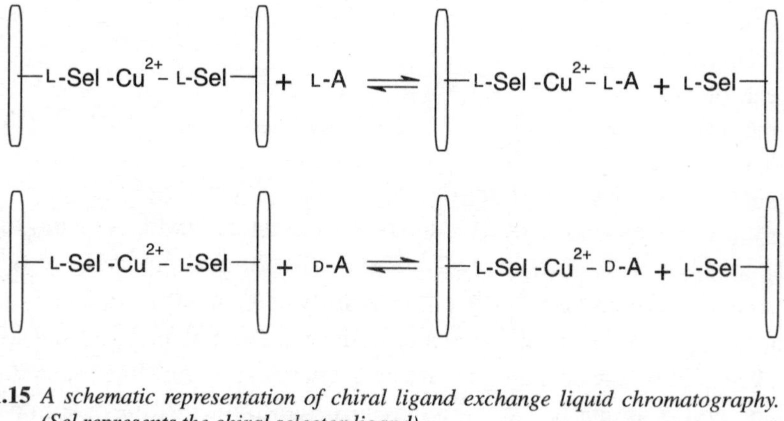
L-Sel -Cu2+- L-Sel
+ L-A
L-Sel -Cu2+- L-A + L-Sel
L-Sel -Cu2+- L-Sel
+ D-A
L-Sel -Cu2+- D-A + L-Sel

Fig. 1.15 *A schematic representation of chiral ligand exchange liquid chromatography. (Sel represents the chiral selector ligand).*

The differential stability of the mixed copper complexes with the L- and D-enantiomers of the amino acids leads to the differences in their retention times and hence the separation. A trace of the chromatograph showing the separation of a pair of enantiomers is given in Fig. 1.16.

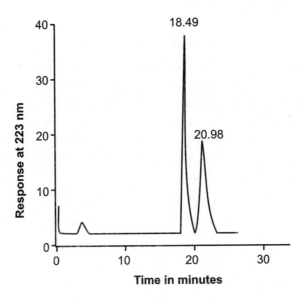

Fig. 1.16 *A representative chiral ligand exchange liquid chromatograph.*

1.8.2 Asymmetric Synthesis of α-Amino Acids

As mentioned earlier, the conventional methods of synthesising α-amino acids give racemic mixtures. Though these can be separated by the procedures outlined above, an ideal situation would be the one where we can prepare optically pure α-amino acids. Nature has evolved selective methods for synthesising optically pure amino acids and also of selectively distinguishing between the enantiomers. In nature, enzymes–the biocatalysts aid in the synthesis of optically pure amino acids from prochiral precursors following a process called **asymmetric synthesis**. It has always been a challenge for a synthetic organic chemist to emulate nature in this respect. Asymmetric synthesis refers to a situation wherein in a reaction sequence at least one reaction is **stereoselective**, intended to preferentially or exclusively yield a particular configuration of the amino acid. In principle, any of the reactions outlined above for the synthesis of α-amino acids can be used for the purpose.

A number of strategies have been successfully worked out to obtain optically pure α-amino acids. These can be broadly put into two categories.

(a) Using chiral catalyst or specific metal ion complexation

(b) Asymmetric induction using chiral auxiliaries

Using chiral catalyst or specific metal ion complexation

A number of synthetic procedures belonging to this class have been evolved. Some of these are detailed below.

(i) **O'Donnell's phase-transfer catalytic alkylation method :** In this method, pseudoenantiomeric quaternary amines derived from the *Cinchona* alkaloid are utilised for the phase-transfer catalytic alkylation of *t*-butylglycinatebenzophenoneimines with alkyl halides. The products obtained can be deprotected with acid to generate α-amino acids in appreciable enantiomeric excess. **Enantiomeric excess** (ee) is defined as given below.

$$\%ee = \frac{R\text{-}enantiomer \ - \ S\text{-}enantiomer}{R\text{-}enantiomer \ + \ S\text{-}enantiomer} \times 100$$

A reaction yielding a mixture of 95 % of *R*-isomer and 5% of *S*-isomer is said to have a 90 % ee.

$$\frac{95-5}{95+5} \times 100 \qquad \Rightarrow \frac{90}{100} \times 100 = 90\%ee.$$

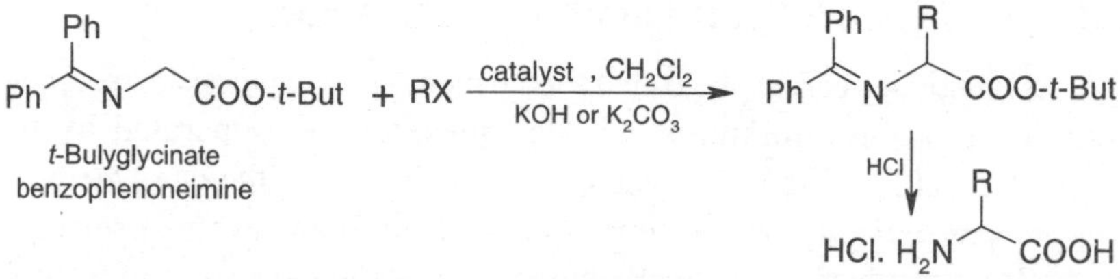

In a specific example, the catalyst used in the reaction mentioned has the following structure.

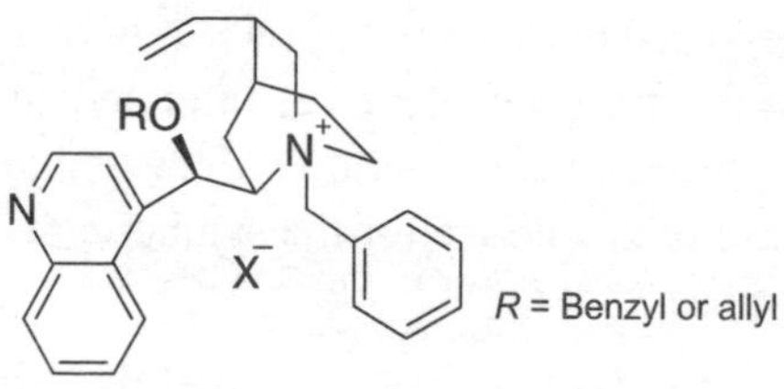

R = Benzyl or allyl

The phase transfer alkylation of *t*-butylglycinatebenzophenoneimine using 4-chlorobenzylchloride as the alkyl halide yielded 4-chloro-L-phenylalanine in 99% enantiomeric excess.

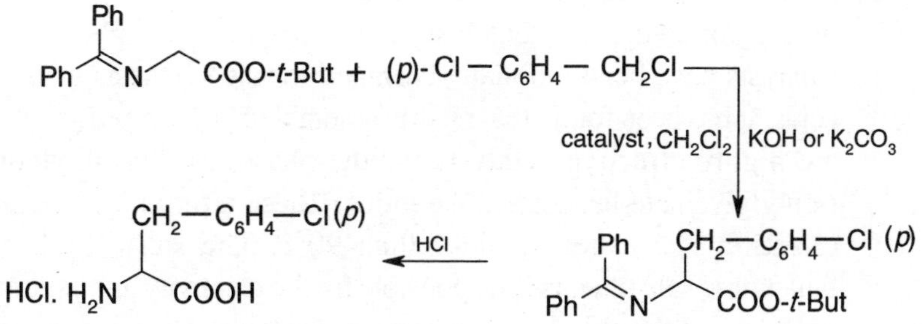

4-Chloro-L-phenylalanine hydrochloride

Minor modifications of the *Cinchona* alkaloid-based quaternary amine phase transfer catalyst can be used to achieve major enhancements in the asymmetric induction. In particular, use of N-9-anthracenylmethyl salts, instead of the N-benzylated adducts used above, generally afforded enantioselectivities greater than 90%.

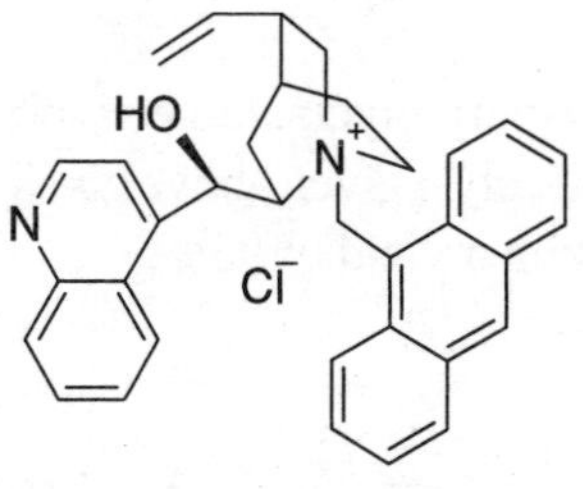

This methodology has been used by Lygo to synthesise α-methyl-α-alkylamino acids from the alanine derived aldimines.

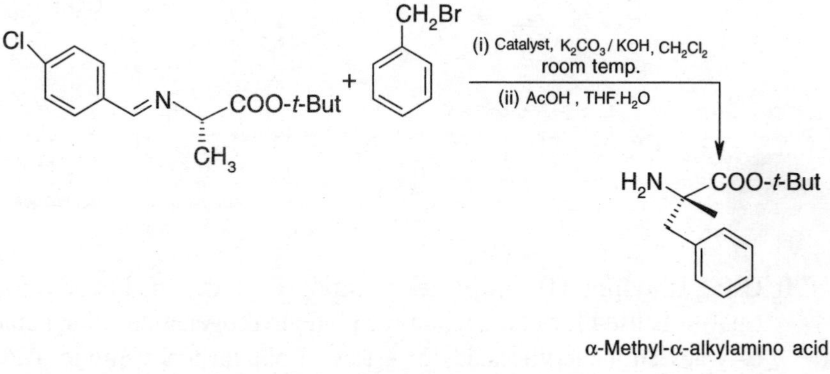

α-Methyl-α-alkylamino acid

(ii) **Non-*Cinchona* alkaloid-derived catalysts :** Following the success of the alkaloid based catalysts a number of novel phase-transfer catalysts have been rationally designed for the synthesis of α-amino acids. It has been found that readily assembled biaryl ammonium salts, are highly effective catalysts for the phase transfer alkylation of *t*-butylglycinate benzophenone imine. These afford high yields and enantiomeric excess greater than 90%. The steric bulk of the binaphthyl substituents is responsible for the reactivity and asymmetric induction of the chiral spiro ammonium catalysts.

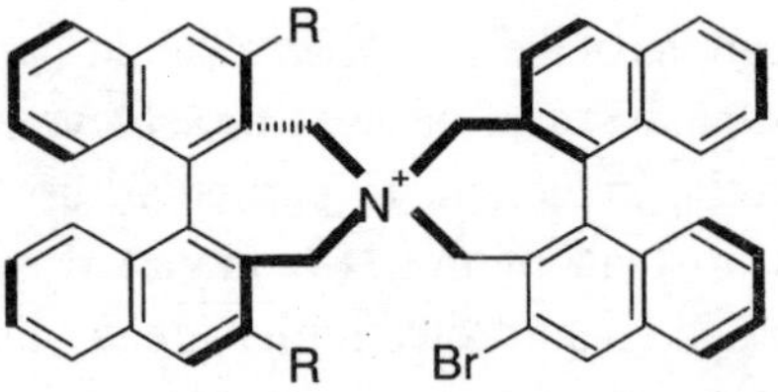

R= β-Naphthyl

The alkylation of *t*-butylglycinate benzophenone imine with benzyl bromide using the catalyst given above gives 96% enantiomeric excess of phenylalanine with a yield of 95%.

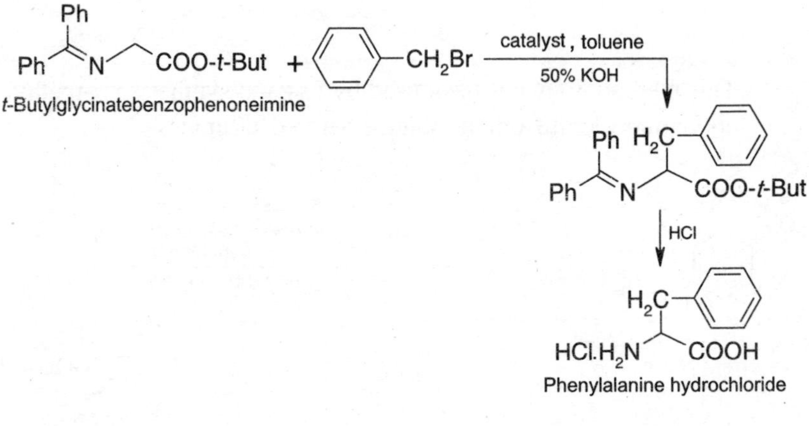

(iii) **Using Rhodium (I) complexes as catalyst :** In this method, a chiral catalyst is used for homogenous catalytic hydrogenation of suitable α-N-acylaminoacrylic acids or esters to obtain N-acylamino acids which on hydrolysis yield the α-amino acid in good enantiomeric excess.

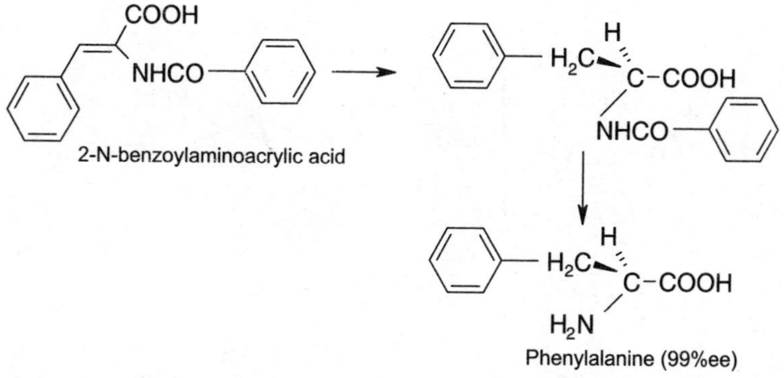

2-N-benzoylaminoacrylic acid

Phenylalanine (99%ee)

In this example, the starting α-N-acylaminoacrylic acid gives the amino acid , phenylalanine, in 99% enantiomeric excess when the reaction is performed in tetrahydrofuran as solvent.

The catalyst used in this reaction is a version of the Wilkinson's catalyst [Rh(Ph$_3$P)$_3$Cl] that is suitably modified to act as an asymmetric catalyst. A chiral ligand called (S,S)-chiraphos [(2S,3S)-bis(diphenylphosphino) butane] is used to chelate the rhodium metal ion to make the desired chiral complex. This chiral complex in presence of hydrogen and a suitable solvent yields the active chiral hydrogenation catalyst. The following scheme represents the formation of the active catalyst.

Synthesis of (S,S)-chiraphos, the ligand

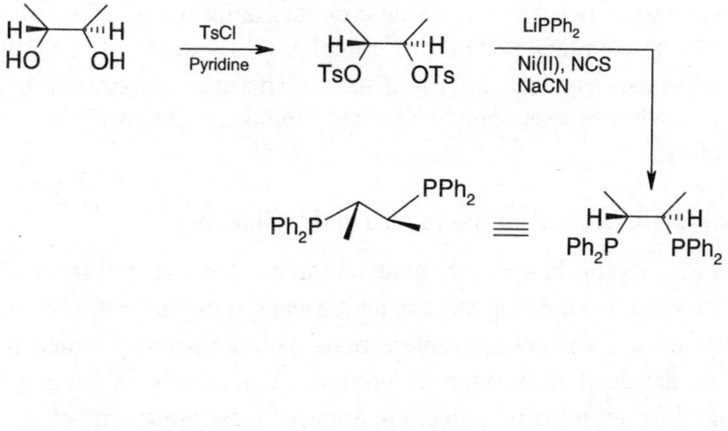

Formation of chiral complex

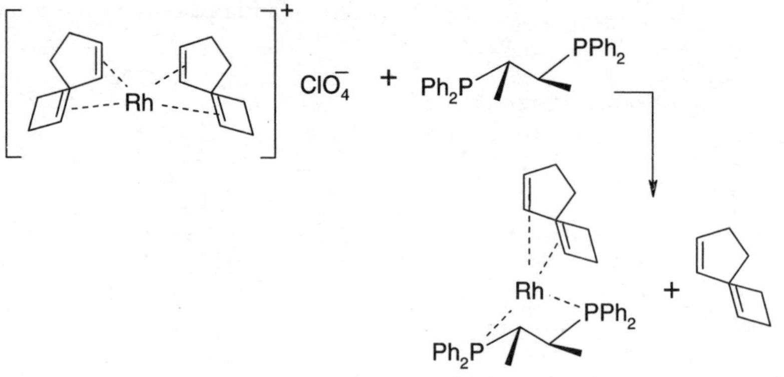

Formation of the active catalyst

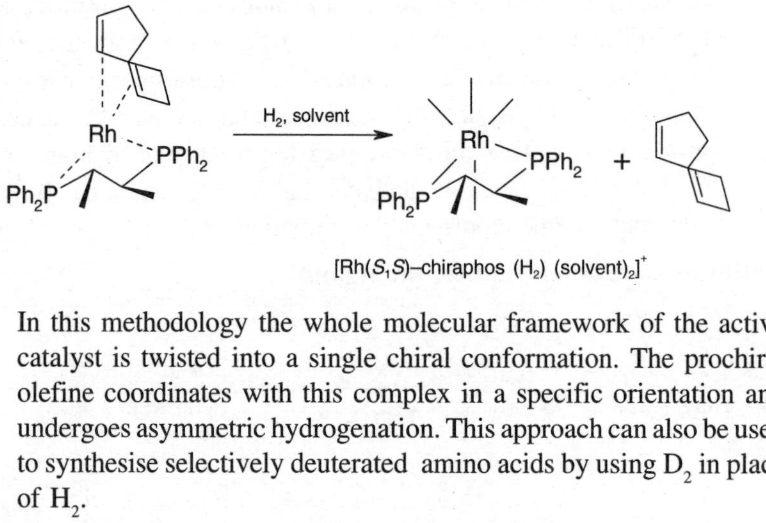

$[Rh(S,S)–chiraphos\ (H_2)\ (solvent)_2]^+$

In this methodology the whole molecular framework of the active catalyst is twisted into a single chiral conformation. The prochiral olefine coordinates with this complex in a specific orientation and undergoes asymmetric hydrogenation. This approach can also be used to synthesise selectively deuterated amino acids by using D_2 in place of H_2.

Asymmetric induction using chiral auxiliaries

In this approach, optically pure substances are prepared from achiral molecules using chiral reagents containing an asymmetric centre. The chiral reagents have diastereotropic interactions with the reactant molecule and generate the desired asymmetric product. A number of strategies have been used for asymmetric induction. Some of these are given below.

(i) Corey's methodology (asymmetric induction) : According to this methodology optically pure α-amino acids are prepared from the corresponding α-keto acids which act as their precursors. It uses a chiral reagent which when combined with the α-keto acid forms a hydrazonolactone ring. The strategy can be outlined as follows:

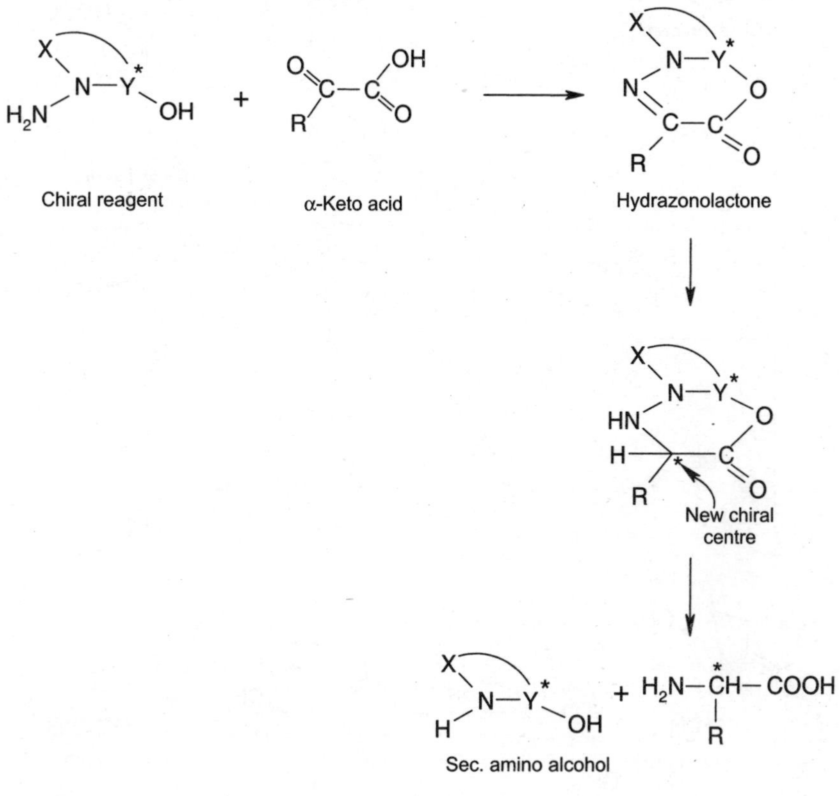

Chiral reagent   α-Keto acid   Hydrazonolactone

New chiral centre

Sec. amino alcohol

The stereospecific reduction of the N = C double bond produces the chiral carbon of the targeted amino acid, which is obtained by hydrogenolysis of the intermediate. The original chiral reagent is regenerated from the chiral secondary amino alcohol obtained.

D-alanine can be synthesised with an optical purity of 80% from methylacetoformate $(CH_3COCOOCH_3)$ using a chiral reagent with a bicyclic indoline structure.

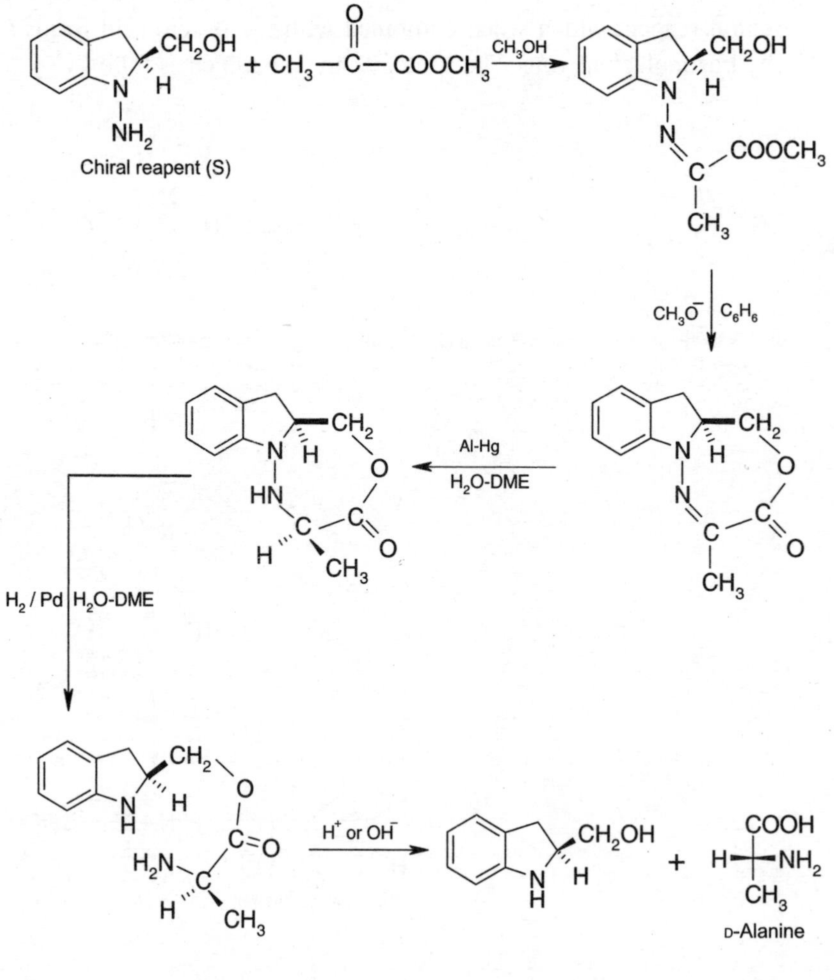

(ii) Using pseudoephedrine as chiral auxiliary : In this method readily available and inexpensive chiral auxiliary, pseudoephedrine reacts with glycine methyl ester to give pseudoephedrine glycinamide methyl ester. The secondary amino group of pseudoephedrine forms an amide bond with the carboxyl group of glycine methyl ester. The alkylation of this substrate with a wide variety of electrophiles proceeds with excellent diastereoselectivity with good yields.

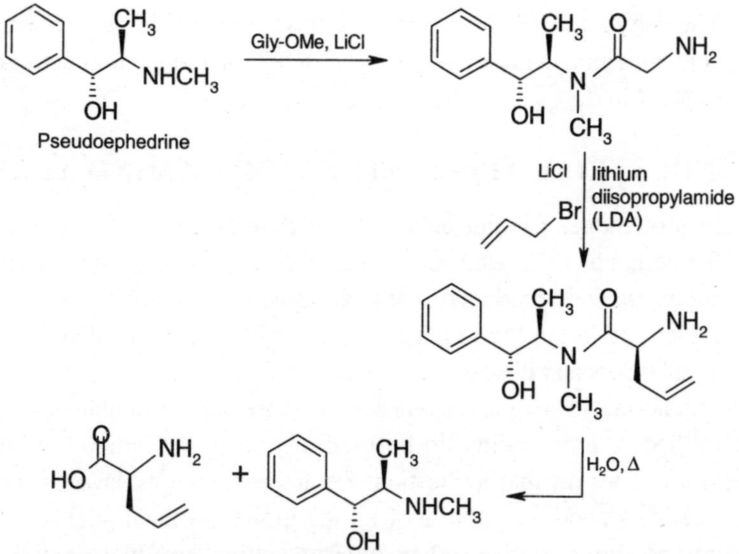

(iii) **Schollkopf's Bis- Lactim approach :** In this method an amino acid, valine, is used as a chiral auxiliary. The auxiliary condenses with glycine to form a bis-lactim ether. The enolisation and electrophillic reaction with an aldehyde or ketone gives an intermediate with high level of distereoface selectivity.

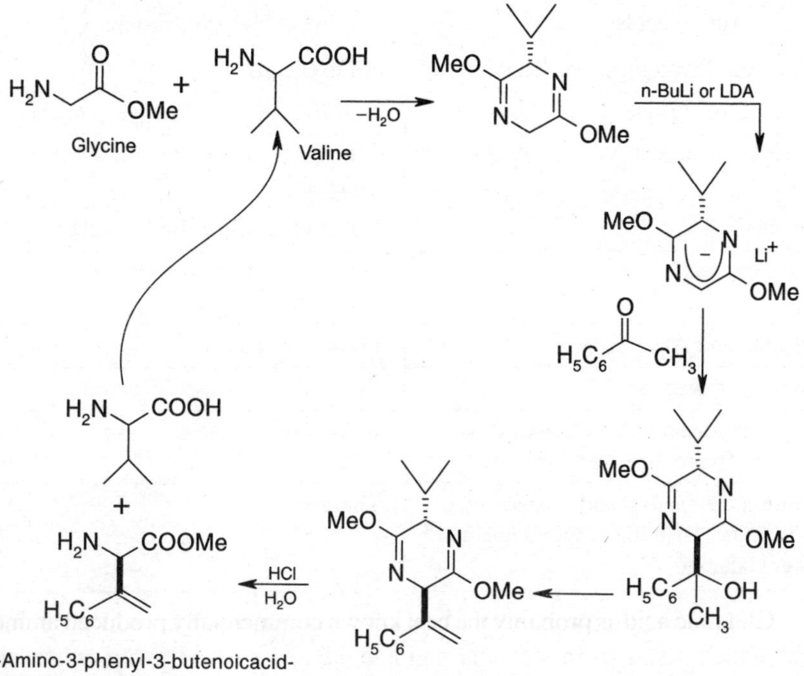

2-Amino-3-phenyl-3-butenoicacid-
methylester (*R*-isomer)

The dehydration and hydrolysis of this intermediate gives the amino acid with an enantiomeric purity of > 95% and the chiral auxiliary is regenerated.

1.9 INDUSTRIAL APPLICATIONS OF α-AMINO ACIDS

The proteinogenic amino acids find varied industrial applications and to meet the demands estimated to be over two million tons, these α-amino acids are extensively produced. These are used in medicine to treat dietary deficiencies and in the food industry as antioxidants, artificial sweeteners, flavour enhancers and in cosmetics. In addition, these are required as cheap starting materials in chemical synthesis and there is a reasonable demand of radiolabelled α-amino acids (to be used as tracers in animal and human studies) also. The principal application of α-amino acids is as flavour enhancer as it consumes about two thirds of all the global production. The annual demand for amino acids used in pharmaceutical products mainly for intravenous and enteral nutrition is to the tune of ~15,000 tons. Table 1.8 lists commercial and industrial applications of some α-amino acids.

Table 1.8 Industrial applications of α-amino acids

Amino acids	Industrial applications
Cysteine, Tryptophan and Histidine	Antioxidants
Methionine, Lysine and Threonine	Nutritive additive in soya products
Phenylalanine and Aspartic acid	Constituents of artificial sweetener aspartame
Lysine	Nutritive additive used in breads
Glutamic acid	Meaty flavoured food additive, meat tenderiser
Glycine and Alanine	Flavour enhancer
Serine and Arginine	Cosmetics
Arginine, Leucine, Isoleucine, Proline Valine, Tryptophan and Tyrosine	Infusions
Arginine, Aspartic acid, Asparagine, Glutamine, Histidine, Methionine and Phenylalanine	Therapy

Glutamic acid is probably the best known commercially produced amino acid which is sold as monosodium glutamate (MSG) –a flavour enhancer. MSG was discovered in 1908 by Dr. Kikunae Ikeda as a basic taste substance of kelp –a traditional seasoning in Japan. Glutamic acid was the first α-amino acid to be produced commercially. It was prepared by extraction from acid hydrolysates of wheat and soya proteins. This was followed by production of a number of amino acids by the same extraction procedure. In late 1950s, fermentation technology was established and used for the commercial production of MSG and marked the beginning of modern amino acid production. This technology was triggered by the discovery of glutamic acid in the exhausted medium of *Cornebacterium glutamicum*. Since then fermentation technologies for a number of amino acids have been established. However, several amino acids like, L-leucine, hydroxy-L-proline, L-tyrosine and L-cystine are still being manufactured by extraction in addition to fermentation and chemical synthesis.

EXERCISES

1 Suppose a new α–amino acid having a –CH_2F group as its side chain is isolated and is named as beucine.

   (a)   Write down the structures of the L-and D- forms of the amino acid beucine.

   (b)   Assign a suitable class to the amino acid.

   (c)   Write the three letter code and suggest a one letter code for beucine.

   (d)   Comment on its optical activity.

   (e)   Ascertain the absolute configuration (R or S) for the α- carbon atom.

2 The next higher homologue (with one more –CH_2 group in the side chain) of beucine (Q 1) is synthesised. Give its structure and assign suitable configuration (in R/S system) to the asymmetric carbon atom.

3 Draw four stereoisomers of the amino acid threonine and assign R/S configurations to the asymmetric carbon atoms.

4 Compute pI values for the amino acids Asp, Val and His (Refer to Table 1.4 for the required pK_a values).

5 Amino acid, serine has a pI value of 5.68. If the carboxyl group has a pK_a value of 2.2, compute the acid dissociation constant for the α-amino group.

6 Using Tables 1.1 and 1.4 write down the structures of the amino acid tyrosine at a pH of

   (a) 2.0   (b) 6.6   (c) 11.0

7 The isoelectric point of glutamine is considerably higher than that of glutamic acid. Explain.

8 A mixture of amino acids contains Asp, Asn, Glu, Gly, His, Lys and Val.

   (a)   Can this mixture be separated by electrophoresis?

   (b)   Predict the order in which these would be obtained on the gel that is buffered at a pH of 6.0.

9 A mixture of amino acids contains glutamic acid, histidine and leucine. Suggest a suitable pH at which electrophoresis of the mixture would give a good separation.

10 Suggest suitable spectroscopic methods that can be used to differentiate between the amino acids alanine and phenylalanine.

11 What differences would you observe in the UV, NMR and CD spectra of the two amino acids mentioned in question 10.

12 Using Tables 1.1 and 1.4, identify

   (a) the most acidic amino acid
   (b) the most basic amino acid
   Justify the choices made.

13 A given sample of alanine produced 1.4 g of N_2 in the Von Slyke determination. Compute the amount of alanine (M_m(Ala) = 89.1 g mol^{-1}) in the given sample.

14 How would you prepare the following α-amino acids using the methods indicated against them

   (a) Valine   Strecker synthesis
   (b) Leucine   Acetamido malonate synthesis
   (c) Isoleucine   Reductive amination
   (d) Tyrosine   Reduction of azlactones
   (e) D-Alanine   Chiral auxiliary
   (f) L-Alanine   4CC synthesis
   (g) L-Phenylalanine   Phthalamido malonic ester synthesis

15 How would you prepare the following α-amino acids using the starting materials indicated against them

   (a) Lysine   1,4-Dibromobutane
   (b) Leucine   Isobutyl alcohol
   (c) Glutamic acid   α-Ketoglutaric acid
   (d) Proline   Adipic acid

16 Give the products formed in the reactions of valine with the following:

   (a) Ethanol; H$^+$
   (b) CH_3COCl; pyridine followed by hydrolysis
   (c) Heating at high temperature
   (d) Phenylisocyanate followed by HCl

17 A synthetic peptide is represented as 'PEPTIDE' using one letter code for amino acids.

   (a) Will it move towards anode or cathode during electrophoresis ?
   (b) Will it absorb UV radiation above 240 nm?

18 Arginine with a guanidine group ($-NH-\overset{\overset{\displaystyle NH}{\|}}{C}-NH_2$) in the side chain is the most basic amino acid. Explain.

19 The pK_a values for β-alanine are 3.55 and 10.24 . Compare these values with that of alanine (cf. Table 1.4) and explain the differences.

20 In strongly acidic solution alanine exists as a mono cation with two potential acidic groups (α-NH_3^+ and $-COOH$).

   (a) Which of these two groups is more acidic?
   (b) How would the structure of alanine change on adding a base?

Chapter 2

Peptides

2.1 INTRODUCTION

Peptides are a large group of biologically active molecules obtained by linking of amino acids through peptide bonds. These are similar to proteins in terms of their composition and mode of formation, the point of distinction being the size. In fact the terms peptide and protein are many a times used interchangeably. The term peptide is suitable if the number of amino acid residues is less than about fifty. Peptides along with proteins exhibit probably the largest structural and functional variation of all classes of biologically active macromolecules in the living systems. Peptides are believed to be involved in biological functions as diverse as regulation of blood pressure, metabolism, thermal regulation, analgesia, enzyme inhibition, sexual maturation and reproduction, learning and memory, etc. These also act as hormones, antibodies, vaccines and toxins.

The study of the structures of bioactive peptides and their relationship with the biological functions along with the developments in the area of peptide synthesis has opened new vistas in preventing and fighting diseases. Since peptides and peptide-related molecules can influence endocrinological, neurological, immunological and enzymatic processes with high specificity and potency, these have varied applications in the field of medicine. The potential areas being regulation of fertility, pain control, growth stimulation, cancer therapy, cardiovascular problems, diseases of connective tissues, digestive disorders, mental illness and infections etc. In fact, several peptide drugs are in widespread clinical use, necessitating their manufacture in large quantities. Synthetic peptides can also be potential candidates for vaccines. A synthetic peptide vaccine avoids the hazards and limitations of making and using vaccines in the conventional way. There has been success in this direction and the future is quite promising.

In this chapter we will take up structure, classification, nomenclature and synthesis of peptides.

2.2   STRUCTURE AND CLASSIFICATION OF PEPTIDES

Two molecules of same or different amino acids condense together with the elimination of a water molecule to form a **peptide**. The reaction may be visualised as the acylation of one amino acid by other amino acid. The amide linkage obtained by the reaction of the carboxyl group of one amino acid to the amino group of another amino acid is called a **peptide bond**. For example, in the following reaction amino acid alanine is acylating valine to give the peptide, alanylvaline. The peptide bond is represented as

$-\overset{\overset{\text{O}}{\|}}{\text{C}}-\text{NH}-$ as shown below.

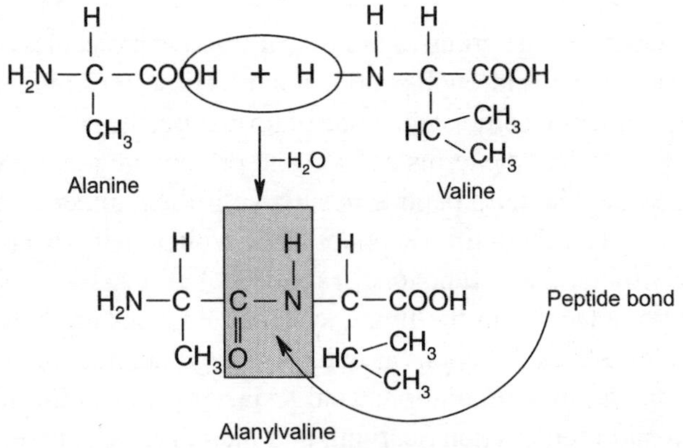

Alanylvaline

More amino acids may condense in a similar way with the free amino or carboxyl group of the peptide to give a larger peptide or a protein.

2.2.1   Structure of Peptide Bond

There are two possible configurations of the planar peptide bond, one having the alpha carbon atoms of the adjacent amino acids *trans* to each other and the other in which these are *cis*. One may visualise *cis* and *trans* in terms of the relative orientations of the C=O and N–H bonds around the C–N bond also. This conformational variable is described in terms of the torsional angle *omega*, which can take values of 0° (*cis*) or 180° (*trans*).

cis   trans

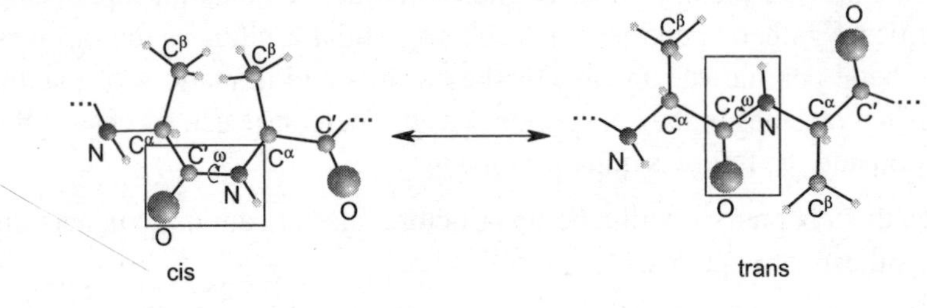

The *trans* form of the peptide is normally the energetically favourable choice as it is estimated to be about 1000 times more stable than the *cis* form. Therefore, in a peptide bond (–CONH–) the hydrogen of the substituted amino group is almost always *trans* to the oxygen of carbonyl group. The bond angles and lengths of a peptide bond, Fig.2.1 (a) are well known from three dimensional crystal structure determinations of a large number of small peptides. The dimensions given in the figure are the average values. The C–N bond length in the peptide bond is observed to be 133 pm, considerably shorter than the non-peptide C–N bond length of 146 pm, but longer than the normal C=N double bond length of 127 pm. This difference is indicative of a partial double bond character of the bond, which can be explained in terms of resonance. The two resonating structures for a *trans* peptide bond can be shown as given in Fig 2.1(b).

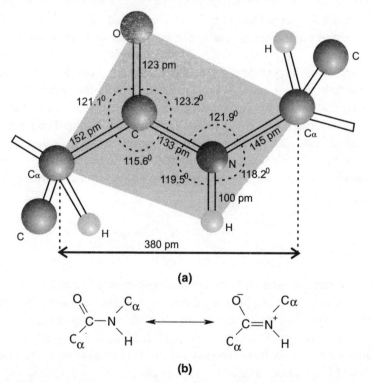

(a)

(b)

Fig. 2.1 *(a) The geometry and the dimensions of a peptide backbone.*
(b) The resonating structures of a trans peptide bond showing partial double bond character of the C–N bond.

It is estimated that the peptide C–N bond has about 40% double bond character which imparts it a kind of rigidity, consequently there is no freedom of rotation about the bond between the carbonyl carbon atom and the nitrogen atom of peptide bond. Due to this, group of six atoms between successive α-carbon atoms is confined to a single plane, Fig.2.1 (a). Thus, a peptide bond is **rigid** and **planar.** The maximum distance between corresponding atoms of adjacent residues is 380 pm when the peptide bond is *trans*. However, in a fully extended polypeptide chain the amino acid residues are staggered and the length of a polypeptide having *n* residues is observed to be (363 x *n*) pm.

In addition to making the peptide bond rigid and planar, the resonance in the peptide bond makes it chemically inert. Since the lone pair of the nitrogen atom enters into resonance, the nitrogen atom is neither basic nor nucleophilic (otherwise expected of an amine). Secondly, the carbonyl group is also stabilised which otherwise can be attacked by a nucleophile. The amide group however, can act as a nucleophile through the carbonyl oxygen.

As can be observed from Fig.2.1 (b), the atoms constituting the peptide bond acquire some charge due to the resonance and the bond becomes polar in nature. The combined effects of polarity and planarity in the *trans* configuration of the peptide bond result in a permanent dipole moment across the peptide bond with its negative end on the side of the carbonyl oxygen. The magnitude of the dipole moment is ~ 3.5 Debye units which is equivalent to unit positive and negative charges separated by 73 pm. The polarity of the peptide bond can be appreciated by comparing it with that of water molecule (1.85 D). The charges acquired by different atoms constituting the peptide bond and the corresponding dipole vector is as shown below.

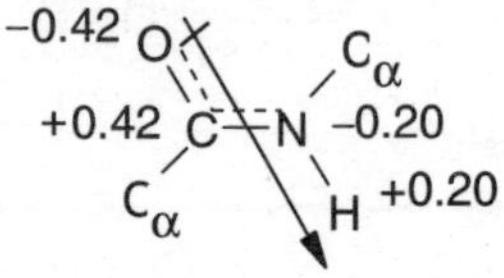

$$-0.42 \quad O \qquad C_\alpha$$
$$+0.42 \quad C-N \quad -0.20$$
$$C_\alpha \qquad H \quad +0.20$$

The charge separation in the peptide bond makes CO and NH potential candidates for H-bonding between different peptide bonds along the peptide backbone. This in turn is responsible for the structures adopted by the peptides and proteins. Planarity of the peptide bond also plays a very crucial role in determining the three dimensional structure of proteins by limiting the number of orientations available to the peptide chain or backbone. On the other hand, the bond between the alpha carbon and the peptide nitrogen atom (145 pm) and the carbon-carbon bond (152 pm) are pure single bonds. There is a large degree of rotational freedom about these bonds on either side of rigid peptide bond.

The *trans* form of the peptide bond, as discussed above, is more stable than the *cis* form. However, if the residue following the peptide bond happens to be proline, the stability of the two forms becomes comparable (4:1). The *cis* configuration is normally found just preceding the proline residue in the peptide or polypeptide chain. As the *cis* configuration affects the configuration of the peptide backbone, the presence of proline residue significantly influences the structure adopted by peptides and proteins.

Chemically, the peptide backbone is not very reactive. The extreme pH values at which the –C=O or NH groups of the peptide bond can undergo significant protonation or deprotonation, the polypeptide gets hydrolysed. However, at physiological pH a peptide bond is quite stable. At neutral pH and room temperature the half-life of a peptide bond is estimated to be about seven years.

2.2.2 Classification of Peptides

The compound formed by the combination of two amino acids through a peptide bond is called a **dipeptide**. If a third amino acid joins the dipeptide by forming a new peptide bond at either the amino or carboxyl terminal (of the dipeptide) a **tripeptide** is obtained and further addition of amino acids will give a **tetra**, a **penta** or a **hexapeptide** and so on. In general, a peptide can be visualised as a short chain of amino acid residues (defined below) that has a defined sequence and generally do not have a fixed three dimensional structure. The designation of the peptide is based on the number of amino acids constituting it. It may be noted here that a dipeptide has one peptide bond while a tripeptide has two and so on. In general the number of peptide bonds in a peptide is one less than the number of amino acid residues in it. A peptide containing 3-10 amino acid residues is called an **oligo-peptide** while the ones containing more than ten residues are referred to as **polypeptides**. The polypeptide backbone is a repitition of the basic unit containing three atoms (amide N, C$^\alpha$ and carbonyl C) of each residue in the chain. The basic repeat unit of the polypeptide can be represented as,

$$\left[\begin{array}{c} \overset{\displaystyle H}{\underset{\displaystyle |}{}} \\ -N-CH-C- \\ \underset{\displaystyle R}{|} \end{array} \overset{\displaystyle O}{\overset{\displaystyle \|}{}} \right]$$

The repeat unit inclusive of the side chain shown in the brackets above is called an **amino acid residue**. Polypeptides with molecular weight of more than 10,000 (50-100 amino acid residues) and having a definite three dimensional structure under physiological conditions are usually termed **proteins**.

2.3   NOMENCLATURE OF PEPTIDES

As mentioned earlier, a peptide formation may be visualised as the acylation of an amino acid by another amino acid accompanied by the elimination of a water molecule. Further elongation of the chain, can similarly be thought of as a consequence of continued acylation of the peptide so obtained. The amino acids, as they exist in the peptides are referred to as amino acid residues. In a peptide chain, the amino acid residues can have structures that lack a hydrogen atom of the amino group (–NH–CHR–COOH), or the hydroxyl moiety of the carboxyl group (NH$_2$–CHR–CO–), or both (–NH–CHR–CO–), R being the side chain. For example, in alanylvaline (Sec. 2.2) the valine residue belongs to the first type while the alanine to the second.

The residue in a peptide that has a free or acylated (but not by an amino acid) amino group is called **N-terminal** residue and the residue with a free or derivatised carboxyl group (as ester or an amide but not acylating other amino acid) is called **C-terminal** residue. The residues are named on the basis of the trivial names of the amino acids. For example, alanine residue, tyrosine residue etc. however, the word 'acid' is omitted while naming aspartic acid and glutamic acid i.e., aspartic residue and glutamic residue respectively.

To construct the name of a peptide, the names of acylating amino acid residues are obtained by replacing '**ine**' ending with '**yl**' while that of the C-terminal residue is used as such. Thus, if the amino acids alanine and valine condense in such a way that alanine acylates valine, the dipeptide formed is named alanylvaline and if they condense in the reverse order, the product is called valylalanine. Structures of some di- and a tripeptide are shown below.

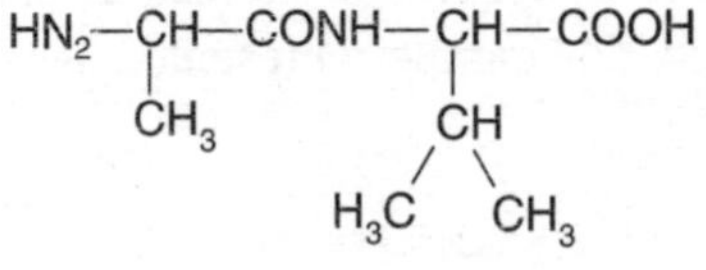

HN$_2$—CH—CONH—CH—COOH   H$_2$N—CH—CONH—CH—COOH
　　　｜　　　　　　｜　　　　　　　　　　　｜　　　　　　｜
　　　CH$_3$　　　　CH　　　　　　　　　CH　　　　CH$_3$
　　　　　　　　／　＼　　　　　　　／　＼
　　　　　　H$_3$C　CH$_3$　　　　H$_3$C　CH$_3$

L-Alanyl-L-valine 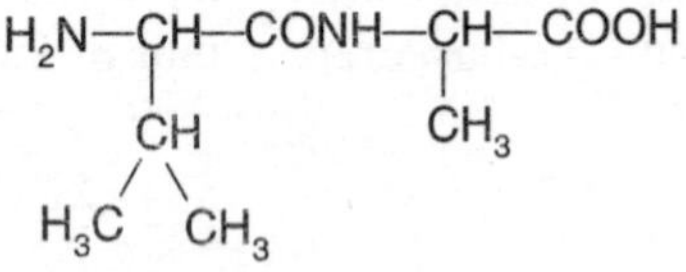   L-Valyl-L-alanine

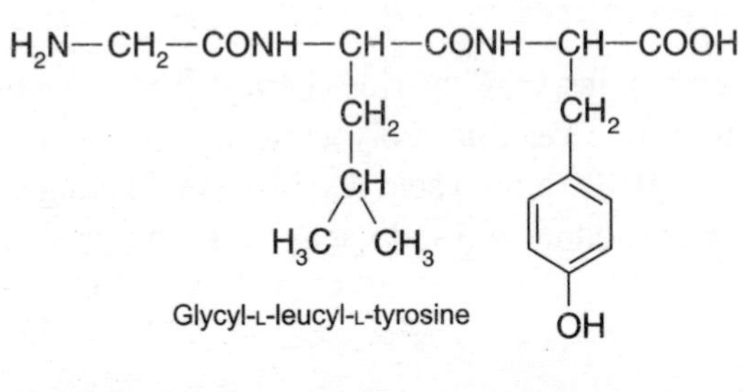

H$_2$N—CH$_2$—CONH—CH—CONH—CH—COOH
　　　　　　　　　　　　｜　　　　　　｜
　　　　　　　　　　　CH$_2$　　　　CH$_2$
　　　　　　　　　　　　｜
　　　　　　　　　　　CH
　　　　　　　　　／　＼
　　　　　　　H$_3$C　CH$_3$

Glycyl-L-leucyl-L-tyrosine   OH

It should be recalled here that all amino acids except glycine are L-amino acids (Sec.1.5). Further, the configurational aspect of different residues is also indicated in the name and is separated from the names before and after them with hyphens. Higher peptides are also named in the same fashion, e.g., glycyl-L-valyl-L-tyrosine represents a tripeptide in which glycine and tyrosine residues are at the N-terminal and C-terminal respectively. Thus the name of the peptide begins with the name of the acyl group representing the N-terminal residue and followed (in order) by the names of the acyl groups representing the internal residues and ends with the name of C-terminal amino acid.

Simple peptides containing a few residues of a certain amino acid are named with prefixes to indicate the number of amino acid residues of the amino acid present, e.g. tetra-L-lysine refers to a peptide containing four residues of L-lysine. If a suitable prefix is not available the corresponding numeral may be used e.g., a 26-peptide refers to a peptide containing twenty-six amino acid residues. Sometimes polymers of amino acids are also prepared. In such cases (as in any polymerisation reaction) we get peptides/polypeptides containing varying numbers of residues. These are given names like oligolysine or poly (L-lysine), etc. depending on the average number of residues (or molar mass). A sample of oligolysine would be a mixture of lysyl peptides with low degree of polymerisation while poly (L-lysine) refers to a sample of lysyl homopolymer with a higher degree of polymerisation. No attempt is made in naming higher oligopeptides and polypeptides of biological origin. These are known by their trivial names and their sequences are described in terms of one or three lettered abbreviated symbols as described below.

2.3.1 Representation of Peptides and Polypeptides

In the formation of a peptide, the amino acid whose carboxyl group participates in the formation of the peptide bond but has an unreacted amino group is called the N-terminal of the peptide and the amino acid with a free carboxyl group is called the C-terminal. While representing the structural formulae of peptides, the N-terminal is conventionally written to the left while the C-terminal is written to the right.

$$
N\text{-terminal } H_2N-\underset{\underset{CH_3}{|}}{CH}-CONH-\underset{\underset{\underset{H_3C\;\;CH_3}{/\;\backslash}}{CH}}{CH}-CONH-CH_2-CONH-\underset{\underset{\underset{\underset{H_3C\;\;CH_3}{/\;\backslash}}{CH}}{CH_2}}{CH}-COOH \;\; C\text{-terminal}
$$

L-Alanyl-L-valyl-glycyl-L-leucine

Even when the formulae of the peptides are written a similar pattern is followed; the N-terminal residue is written on the left, and the C-terminal on the right, e.g.

$$NH_2-CH(CH_2COOH)-CO-NH-CH(CH_3)-CO-NH-CH_2-COOH$$

L-Aspartyl-L-alanylglycine

The complete order of amino acids in a protein is called its **sequence** and is conveniently expressed by using the abbreviated names of the amino acids read from N- to C- terminal. The presence of peptide bonds between different residues is indicated by hyphens. The thirty residue long, amino acid sequence of the B chain of insulin in terms of three letter abbreviations is represented as

Phe-Val-Asn-Gln-His-Leu-Cys-Gly-Ser-His-Leu-Val-Glu-Ala-Leu-Tyr-leu-Val-Cys-Gly-Glu-Arg-Gly-Phe-Phe-Tyr-Thr-Pro-Lys-Ala.

In terms of one letter abbreviation it is given as

FVNOQHLCGSHLVEAYLCGERGFFYTPKA

The nintyneine amino acid residues of HIV *protease* are in the following sequence.

(N-terminal) Pro-Gln-Ile-Leu-Trp-Gln-Arg-Pro-Leu-Val-Thr-Ile-Lys-Ile-Gly-Gly-Gln-Leu-Lys-Glu-Ala-Leu-Leu-Asp-Thr-Gly-Ala-Asp-Asp-Thr-Val-Leu-Glu-Glu-Met-Ser-Leu-Pro-Gly-Arg-Trp-Lys-Pro-Lys-Met-Ile-Gly -Gly-Ile-Gly-Gly-Phe-Ile-Lys-Val-Arg-Gln-Tyr-Asp-Gln-Ile-Leu-Ile-Glu- Ile-Cys-Gly-His-Lys-Ala-Ile-Gly-Thr-Val-Leu-Val-Gly-Pro-Thr-Pro-Val-Asn-Ile-Ile-Gly-Arg-Asn-Leu-Leu-Thr-Gln-Ile-Gly-Cys-Thr-Leu-Asn-Phe (C-terminal)

2.4   PEPTIDE SYNTHESIS

The first synthetic peptide, benzoyl diglycine was prepared accidentally by Theodor Curtius in 1881. However, the synthetic peptide era started in the beginning of the twentieth century when Emil Fischer suggested and tried different strategies to make small peptides. On the basis of analysis of the hydrolysis products of proteins, he suggested that proteins are made up of a large number of α-amino acids linked together by amide linkages called peptide linkages. He supported this conclusion by actually synthesising a peptide, glycylalanine. This was done by first reacting α-chloroacetylchloride with the amino group of alanine ester and hydrolysing the product so obtained as per the following reaction sequence.

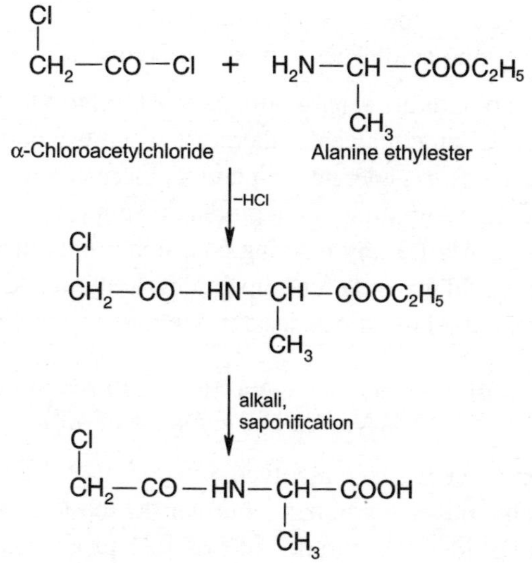

$$\underset{\alpha\text{-Chloroacetylchloride}}{\overset{\overset{\displaystyle Cl}{|}}{CH_2-CO-Cl}} \quad + \quad \underset{\text{Alanine ethylester}}{H_2N-\underset{\underset{\displaystyle CH_3}{|}}{CH}-COOC_2H_5}$$

$$\downarrow \text{-HCl}$$

$$\underset{\underset{\displaystyle CH_3}{|}}{\overset{\overset{\displaystyle Cl}{|}}{CH_2}-CO-HN-CH-COOC_2H_5}$$

$$\downarrow \text{alkali, saponification}$$

$$\underset{\underset{\displaystyle CH_3}{|}}{\overset{\overset{\displaystyle Cl}{|}}{CH_2}-CO-HN-CH-COOH}$$

The chloro substituent in free acid so obtained was then converted to an amine by treating it with aqueous ammonia for several days finally to obtain the peptide.

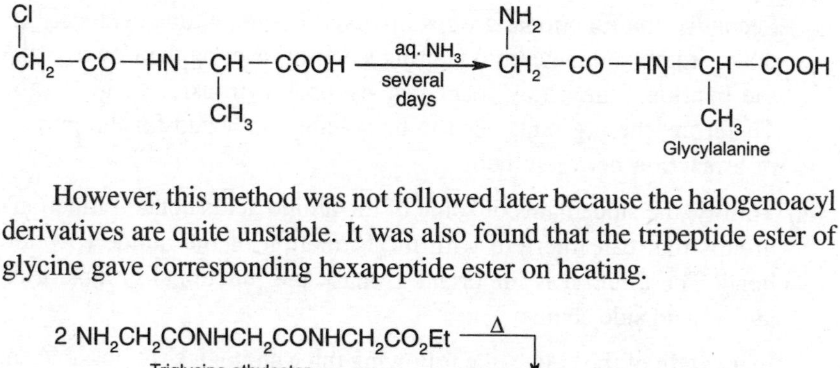

$$\overset{\overset{\displaystyle Cl}{|}}{CH_2}-CO-HN-\underset{\underset{\displaystyle CH_3}{|}}{CH}-COOH \xrightarrow[\substack{\text{several}\\ \text{days}}]{\text{aq. NH}_3} \underset{\text{Glycylalanine}}{\overset{\overset{\displaystyle NH_2}{|}}{CH_2}-CO-HN-\underset{\underset{\displaystyle CH_3}{|}}{CH}-COOH}$$

However, this method was not followed later because the halogenoacyl derivatives are quite unstable. It was also found that the tripeptide ester of glycine gave corresponding hexapeptide ester on heating.

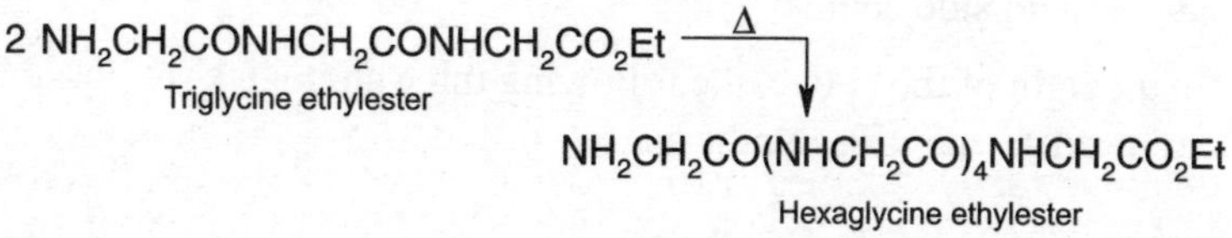

$$2\ \underset{\text{Triglycine ethylester}}{NH_2CH_2CONHCH_2CONHCH_2CO_2Et} \xrightarrow{\Delta}$$

$$\underset{\text{Hexaglycine ethylester}}{NH_2CH_2CO(NHCH_2CO)_4NHCH_2CO_2Et}$$

Nonetheless this strategy could be used to synthesise an octadecapeptide (a peptide containing eighteen residues). The 'art' of synthesising peptides has evolved a great deal since then. Before taking up the strategies of peptide synthesis let us try to understand the 'special' requirements of peptide synthesis. The formation of an amide linkage involves the reaction of an activated carboxylic acid with an amine and a peptide is obtained by formation of an amide linkage between the carboxyl group of one amino acid and the amino group of the other. It appears to be quite simple and straightforward, however, there are a number of practical difficulties in the process of making a peptide from amino acid. These are stated below.

(i) Firstly, amino acids are at least bifunctional (amino and carboxyl groups are part of the same molecule) therefore when we attempt to combine two amino acids to generate a dipeptide, these can react in more than one way to give a mixture of the products. Suppose we wish to prepare the dipeptide Ala-Leu by reacting equimolar amounts of alanine and leucine. It would generate four dipeptides, Leu-Ala, Ala-Ala, Leu-Leu and the desired Ala-Leu besides a mixture of tri and higher peptides.

$$Ala + Leu \longrightarrow Ala\text{-}Leu + Ala\text{-}Ala + Leu\text{-}Ala + Leu\text{-}Leu$$
$$+ Ala\text{-}Ala\text{-}Leu + Ala\text{-}Leu\text{-}Leu + \cdots$$

In addition to the wastage involved, separation of these closely related peptides also poses a problem. Further, as the length of the peptide increases so does the number of possible products. In the case of tripeptides, the number of possible tripeptides from these two amino acids rises to eight. Therefore, in order to synthesise a peptide from its component amino acids, this statistical factor must be overcome i.e., some kind of selectivity must be invoked to get the desired product.

(ii) Secondly, the functional groups involved in the peptide synthesis i.e., the α-amino and the carboxyl groups do not react on their own to give the peptide. They may react just by proton transfer to give salts. Therefore, these groups need to be suitably activated for the purpose of a reaction between them.

(iii) Thirdly, the side chains of some of the amino acids contain functional groups that can interfere with the formation of the amide (peptide) bond. Therefore, it is important to mask the functional groups of the amino acid side chains.

In the light of the above, the following three strategies are possible and are used for peptide synthesis.

Strategies for Peptide Synthesis

(I) *Protecting the amino group and activating the carboxyl group of the first amino acid (aa-1) and coupling (combining) with the second amino acid (aa-2).* The protection of the amino group deactivates it and ensures that the amino group of only the second amino acid reacts to generate the desired peptide. Here 'P' refers to the protecting group while 'A' indicates an activating group.

This would give a dipeptide that is protected at the N-terminal. The peptide chain can be extended by activating the carboxyl group of the dipeptide and combining with another amino acid and continuing the process further. The final product is then deprotected in a way that it does not affect the peptide bond to get the free peptide, which is then appropriately purified.

(II) *Protecting the amino group of the first amino acid and the carboxyl group of the second amino acid and combining the two using a suitable coupling method.*

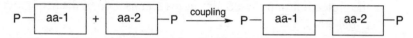

This would give a dipeptide that is protected at both the ends. The peptide chain can be extended by deprotecting the carboxyl group of the dipeptide and combining with another amino acid protected at the carboxyl end or deprotecting the amino group and combining with a N-protected amino acid and continuing the process. The requirement is to deprotect one of the protecting groups without affecting the other. The final product is then deprotected at both the ends to get the free peptide. It is then appropriately purified.

(III) *Protecting the amino group and activating the carboxyl group of the first amino acid and coupling it with the second amino acid with its protected carboxyl group.*

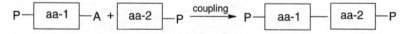

This would also give a dipeptide that is protected at both the ends. The peptide chain can be extended by deprotecting the carboxyl group of the dipeptide, activating it and combining with another amino acid protected at C-terminal and continuing the process. The final product is then deprotected at both the ends to get the free peptide.

It is implicit that the side chain functional groups that can interfere in the process of peptide bond formation are also suitably protected during the process of peptide bond formation. These protecting groups are also deprotected along with the N- and C-terminal protecting groups.

2.4.1 Protection of Amino Group

As discussed above, in order to be able to make a peptide bond between the amino acids in the desired sequence, we need to deactivate the amino functional group of the N-terminal amino acid. It is accomplished by the acylation of amino group by treatment with acyl chlorides or anhydrides. This acyl (amide) formation substantially reduces the basicity and nucleophilicity of amines whereby the amino groups no longer can participate in the peptide bond formation. The following points need to be kept in mind while designing an amino protecting group.

- the protecting group should be easy to introduce.
- the protected amino group should not react under peptide forming conditions (coupling reactions).
- the protection step should not cause racemisation of the neighbouring chiral atom.
- the protected amide group formed as a result of protection should be easy to remove under mild conditions without harming the new peptide bond.

A number of N-protecting groups that satisfy these conditions have been devised. We shall take up the better known ones.

(i) Benzyloxycarbonyl (Cbz) Group

The benzyloxycarbonyl group earlier known as carbobenzoxy (Cbz or simply Z group) was introduced by M. Bergmann and Zervas in 1932. It is introduced by reaction of amino acid with benzylchloroformate in aqueous organic solvent mixtures under alkaline conditions. The pH is kept sufficiently high so as to ensure a good fraction of unprotonated amino group.

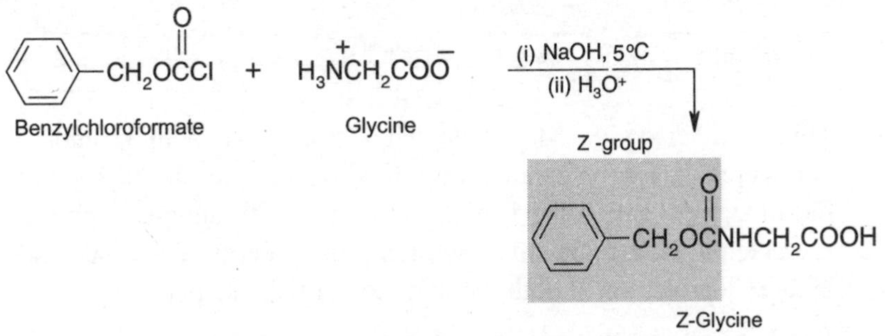

Benzylchloroformate   Glycine

Z -group

Z-Glycine

As the reaction proceeds more base is required because the HCl formed in the reaction lowers the pH. The reaction can be performed in a pH-stat so that the pH is maintained during the reaction. The benzylchloroformate reagent is prepared by treating benzyl alcohol with phosgene.

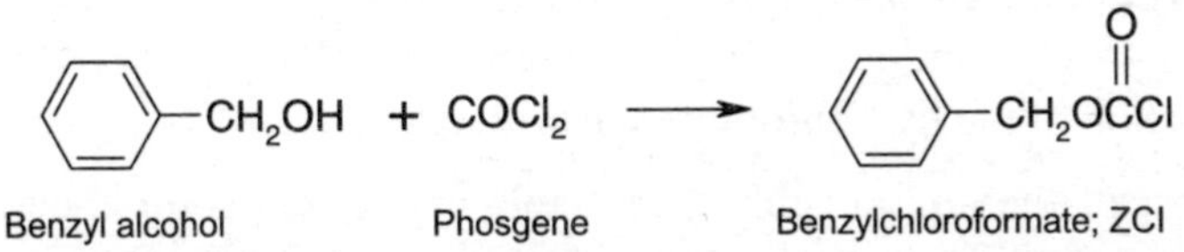

Benzyl alcohol   Phosgene   Benzylchloroformate; ZCl

The benzyloxycarbonyl group is normally not affected by mildly basic or nucleophilic reagents like amines, hydrazine etc at room temperature. It is stable to trifluoroacetic acid (TFA) in ice bath for about an hour. However, it can be removed easily by catalytic hydrogenation using palladium charcoal catalyst at room temperature and ordinary pressure. In this hydrogenation reaction the benzyl-oxygen bond is cleaved to generate carbamic acid that decarboxylates to give the amino acid with free amino group. The poisoning of the catalyst by divalent sulphur from methionine or cysteine may pose a problem in this procedure.

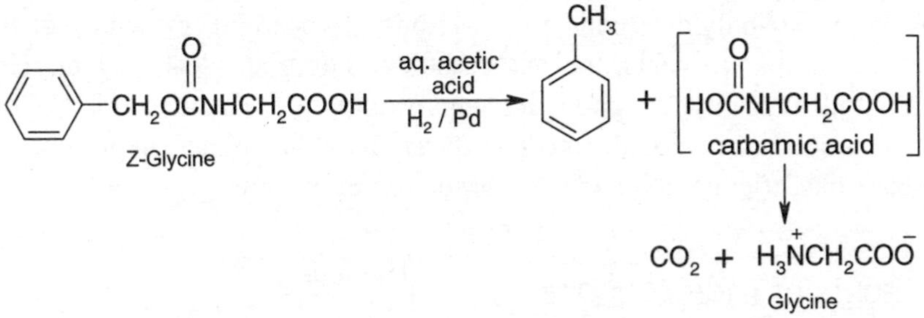

Z-Glycine

The deprotection can also be achieved by using sodium in liquid ammonia or anhydrous acids like hydrobromic acid in glacial acetic acid or HF.

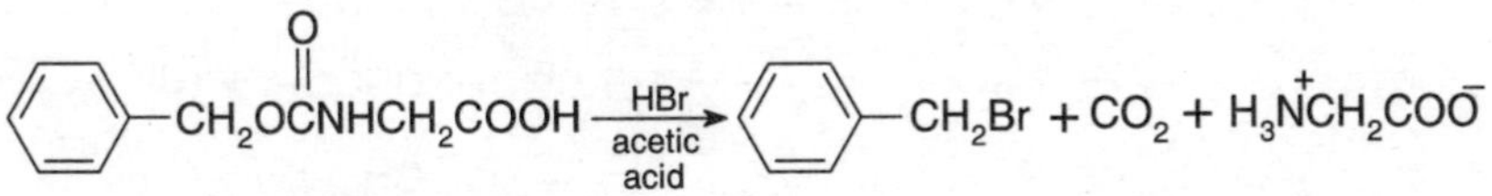

The cleavage of the Cbz group by HBr / acetic acid proceeds by S_N2 mechanism as shown below.

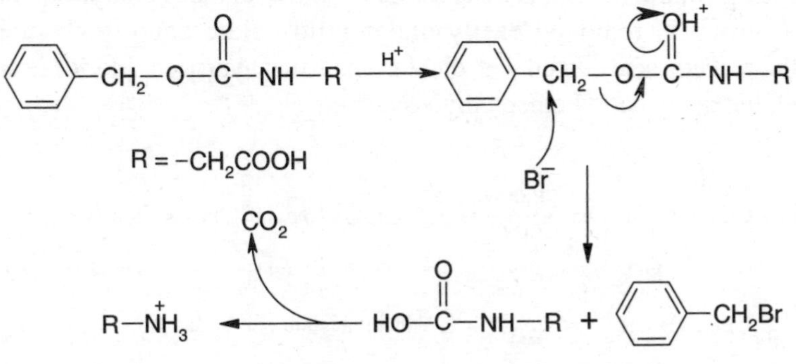

$R = -CH_2COOH$

It is quite a convenient procedure due to good solubility of the protected peptide in acetic acid especially in presence of HBr and the possibility of precipitation of the hydrobromide salt with ether.

Most of the amino acids give stable crystalline solid Z-derivatives. However, when the crystallisation of the derivatives is not easy, certain structural variations of the Z group are used. The commonly used variants are, p-chlorobenzyloxycarbonyl, Z(Cl) and p-nitrobenzyloxycarbonyl, Z (NO$_2$) group. These are also deprotected the same way as the Z-group. Though a useful protecting group, this classical group (Z) has not found widespread use in solid phase peptide synthesis (SPPS) discussed later.

(ii) t-Butoxycarbonyl (Boc or t-Boc) Group

t-Butoxycarbonyl group (Boc) is one of the most popular α-amino protecting groups. Boc group can be introduced easily using t-butoxycarbonyl azide, di-*tert*-butyldicarbonate (di-*tert*-butylpyrocarbonate) or any of the several similar reagents, the pyrocarbonate being an ideal reagent. This anhydride reagent gives good yield and is quite safe. The nucleophilic acyl substitution reaction of alanine with di-*tert*-butyldicarbonate in presence of a base like, triethylamine (TEA) can be represented as:

$$
\underset{\substack{\text{Alanine}}}{\overset{\displaystyle O}{\underset{\displaystyle CH_3}{H_2NCH\overset{\displaystyle \|}{C}OH}}} + \underset{\text{di-\textit{tert}-butyldicarbonate}}{(CH_3)_3C\overset{\displaystyle O}{\overset{\displaystyle \|}{C}}OCOC(CH_3)_3} \xrightarrow{(CH_3CH_2)_3N}
$$

Boc-group

$$
\underset{\substack{\text{Boc-Alanine}}}{(CH_3)_3CO\overset{\displaystyle O}{\overset{\displaystyle \|}{C}}HNCH\overset{\displaystyle O}{\overset{\displaystyle \|}{C}}OH} + (CH_3)_3COH
$$

Unlike the Z group, Boc group is stable to hydrogenation and general reducing conditions and is more labile to acid than the Z group. Therefore, Boc group is removed easily under mild acidic conditions such as trifluoroacetic acid diluted with dichloromethane; hydrogen chloride in acetic acid, dioxane, diethyl ether, ethylacetate etc.

$$
\underset{\substack{\text{Boc-Alanine}}}{(CH_3)_3CO\overset{\displaystyle O}{\overset{\displaystyle \|}{C}}HNCH\overset{\displaystyle O}{\overset{\displaystyle \|}{C}}OH} \xrightarrow[CH_2Cl_2]{CF_3COOH} \underset{\substack{\text{Trifluoroacetate salt of alanine}}}{CF_3CO\bar{O}\ H_3\overset{+}{N}CH\overset{\displaystyle O}{\overset{\displaystyle \|}{C}}OH} + CO_2 + (CH_3)_2C{=}CH_2
$$

The mechanism of deprotection is similar to the one in case of Z group. The cleavage of benzyl-oxygen bond generates carbamic acid that decarboxylates to give the amino acid with free amino group.

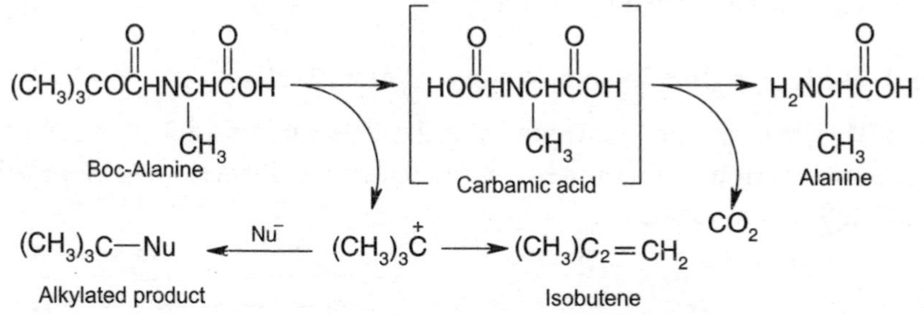

The *t*-butyl carbocation formed in the process either gives isobutene or is trapped by a nucleophile else it may lead to the undesired alkylation of the side chain of tryptophan and methionine (if present). Therefore, in such a situation certain thiophenols or thioethers which can be easily alkylated by the carbocation, are used during the deprotection step.

Since the Boc group is completely stable towards catalytic hydrogenation, it can be used when the Z group is used for the side chain protection. The two groups can be removed selectively in presence of each other and are said to be **orthogonal**.

(iii) 2-(4-Biphenylyl)-isopropoxycarbonyl (Bpoc) Group

This group is similar to the Boc group but is more acid labile than Boc due to the biphenylyl structure, which stabilises the carbonium ion. Bpoc group is introduced with the help of Bpoc-azide (2-(4-biphenylyl)-isopropoxycarbonyl azide)

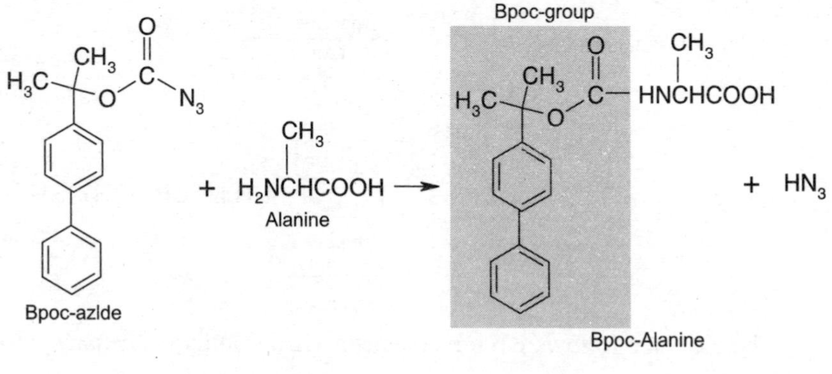

The Bpoc group can be deprotected by treating with chloroacetic acid in dichloromethane. The Boc and Z group remain unaffected under these conditions. Therefore, the Bpoc group can be used along with these groups and can be selectively removed, if required. However, like Z group it can be cleaved by catalytic hydrogenation.

(iv) 9-Fluorenylmethyloxycarbonyl (Fmoc) Group

This group is introduced by reacting the amino acid in a Schotten Baumen type of reaction with fluoren-9-ylmethylchloroformate or fluoren-9-ylmethylcarbonyl azide.

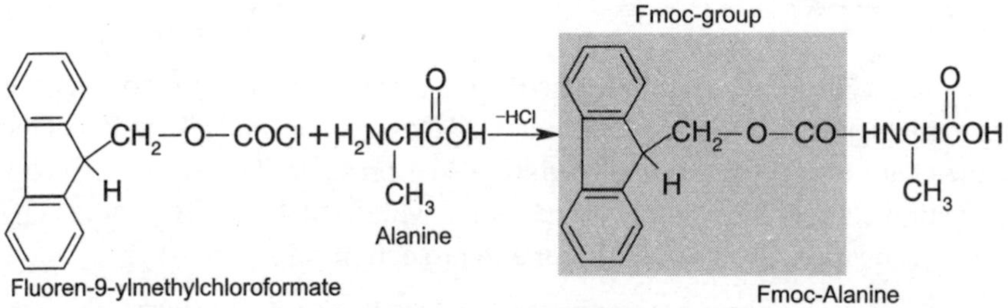

Nowadays a less reactive reagent, FmocOsu-, the succimido ester is becoming popular as it gives better results than the chloroformate. Fmoc is quite stable to acidic deprotection conditions used for removing Z and Boc groups but is quite labile to bases like morpholine and piperidine (a cyclic secondary amine). Fmoc group is cleaved under mildly basic, nonhydrolytic conditions. A 20% solution of piperidine in DMF is the reagent of choice and cleaves the Fmoc group completely in about two minutes.

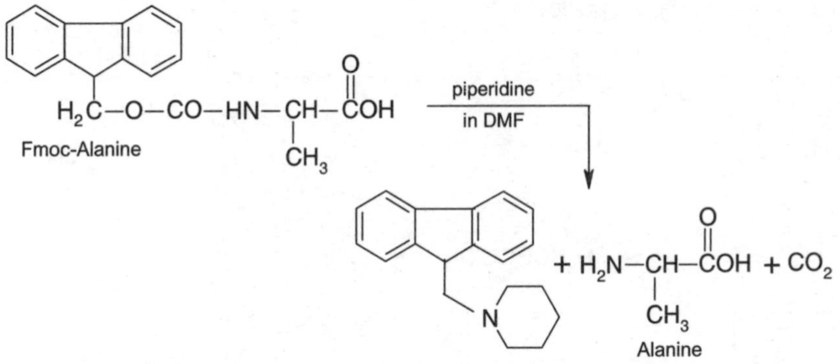

The cleavage follows E1cb mechanism via a stabilised dibenzocyclopentadienide anion. The dibenzofulvene so produced reacts with piperidine to give the co-product.

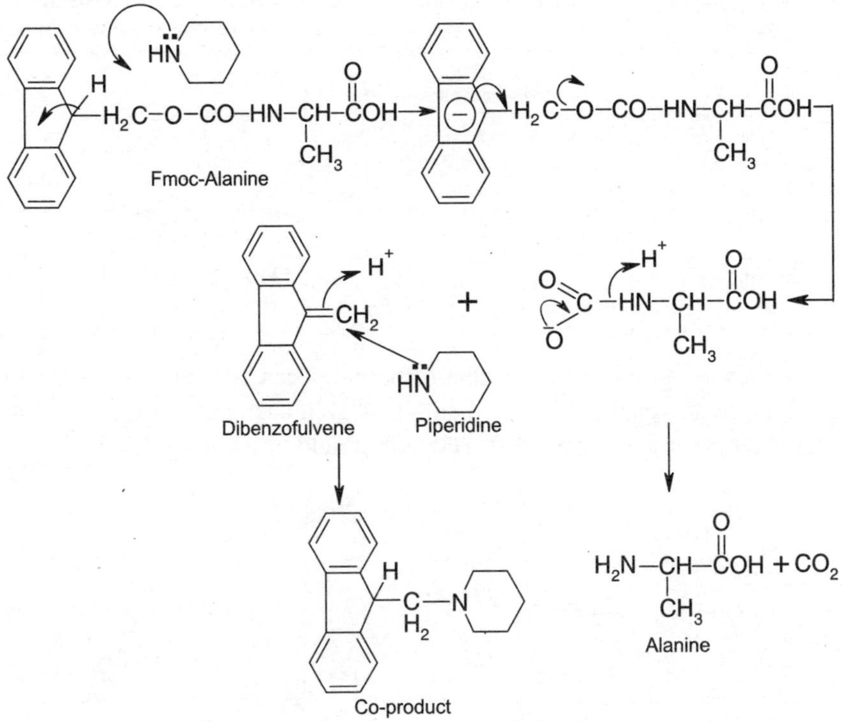

Fmoc-Alanine

Dibenzofulvene Piperidine

$H_2N-CH-\overset{O}{C}OH + CO_2$
CH₃
Alanine

Co-product

This makes Fmoc successful as a protecting group because during its deprotection the peptide bond and *t*-butyl protected side-chain ester linkages are largely unaffected.

(v) Trityl (Triphenylmethyl) Group

It is introduced by reacting the amino acid with triphenylmethyl (or trityl) chloride in an organic solvent containing a base like triethylamine (Et₃N).

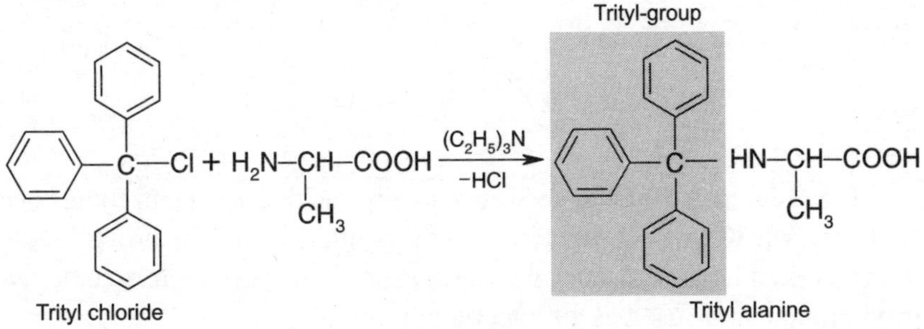

Trityl chloride

Trityl-group

Trityl alanine

Though convenient, the yields are quite low in this method. A modification of this procedure involves a reaction of the amino acid with trimethylsilylchloride in presence of Et_3N followed by treatment of the adduct with trityl chloride.

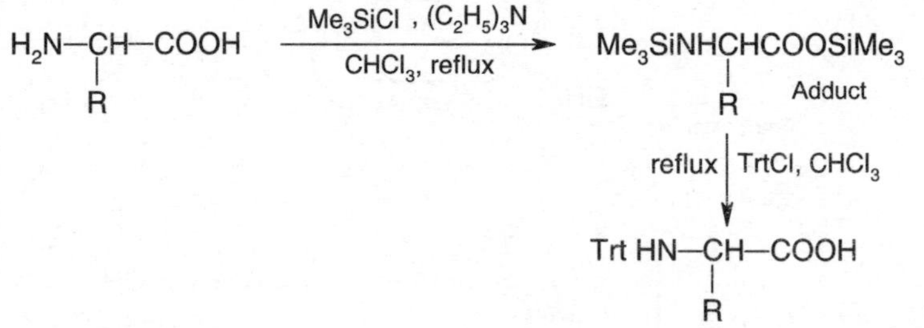
$H_2N-CH-COOH$
R
Me_3SiCl , $(C_2H_5)_3N$
$CHCl_3$, reflux
$Me_3SiNHCHCOOSiMe_3$
R
Adduct
reflux
TrtCl, $CHCl_3$
Trt HN—CH—COOH
R

The trityl group is very stable to base, but is quite labile to acid and catalytic hydrogenation. It is so labile that even acetic acid can remove it. The trityl group is cleaved by H_2/Pd, acetic acid or trifluoroacetic acid.

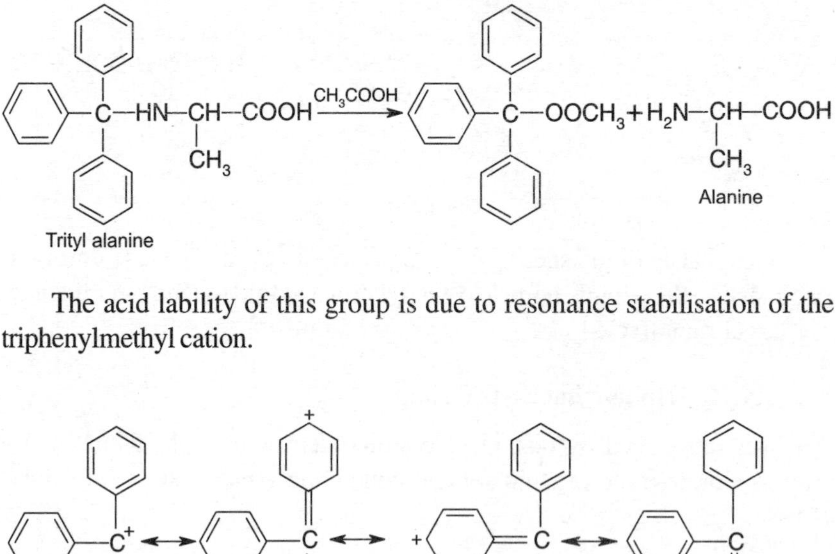
C—HN—CH—COOH
CH_3
CH_3COOH
C—$OOCH_3$+H_2N—CH—COOH
CH_3
Alanine

Trityl alanine

The acid lability of this group is due to resonance stabilisation of the triphenylmethyl cation.

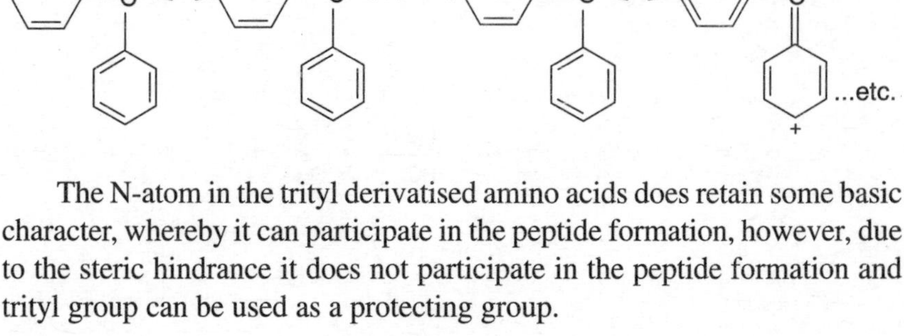
C^+
C
C
C
...etc.

The N-atom in the trityl derivatised amino acids does retain some basic character, whereby it can participate in the peptide formation, however, due to the steric hindrance it does not participate in the peptide formation and trityl group can be used as a protecting group.

(vi) Nitrophenylsulphenyl (Nps) Group

It is introduced by reacting the amino acid with 2-nitrosulphonylchloride under basic conditions.

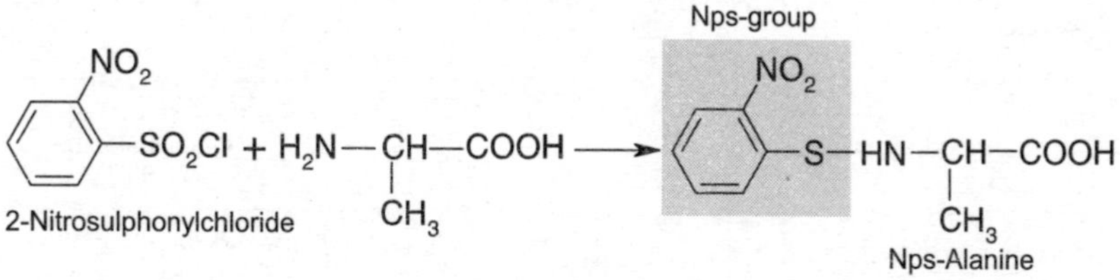

The Nps derivatives are not quite stable in the free acid form. Therefore, these are isolated, purified and stored as dicyclohexylamine salts. The free acid derivative can be liberated from the salt by careful acidification with sulphuric acid. The Nps group is stable to mild base but cannot withstand acidic or hydrogenolysis conditions. The deprotection is carried out with the help of two equivalents of anhydrous hydrogen chloride in an inert solvent.

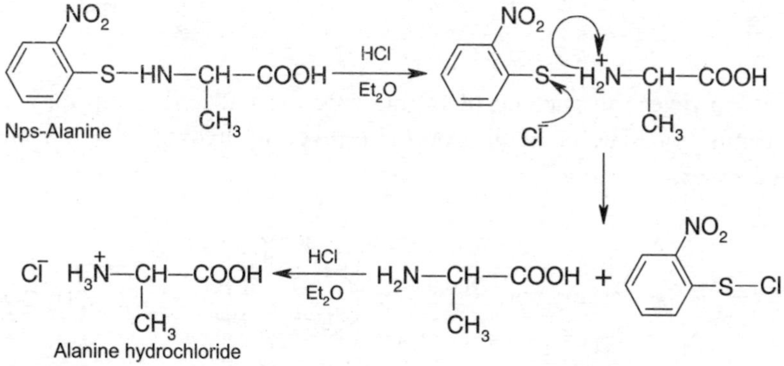

(vii) Phthalyl Group

It is introduced by the reaction of amino acid with N-carbethoxy-phthalimide, which in turn is prepared from potassium phthalimide and ethylchloroformate.

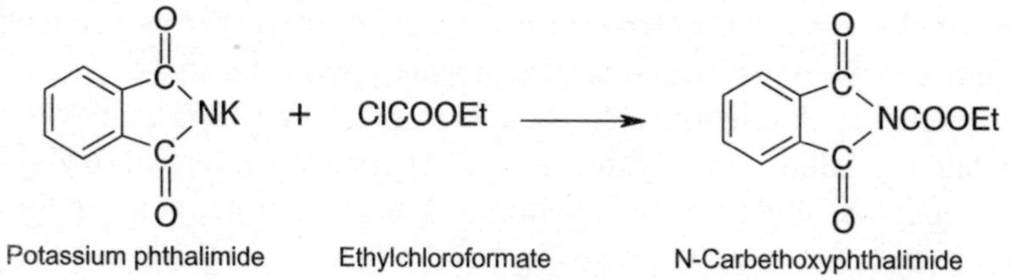

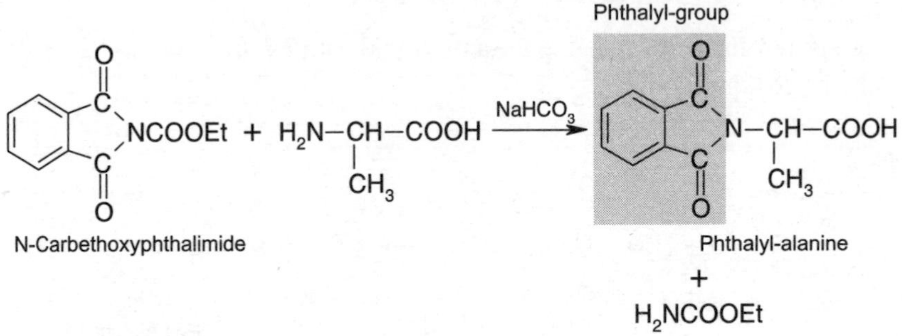

Phthalyl-group

N-Carbethoxyphthalimide   Phthalyl-alanine
                                                                    +
                                                            H$_2$NCOOEt

The phthalyl group may also be introduced directly by the reaction of amino acid with phthalic anhydride.

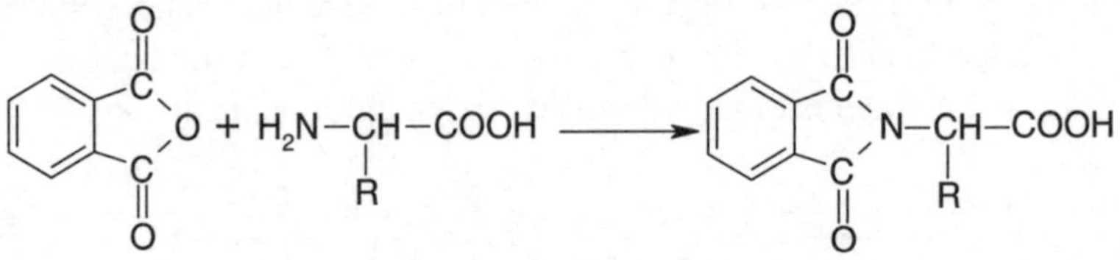

However this method of introducing the phthalyl group may cause racemisation. The phthalyl group is cleaved by hydrazine or substituted hydrazine.

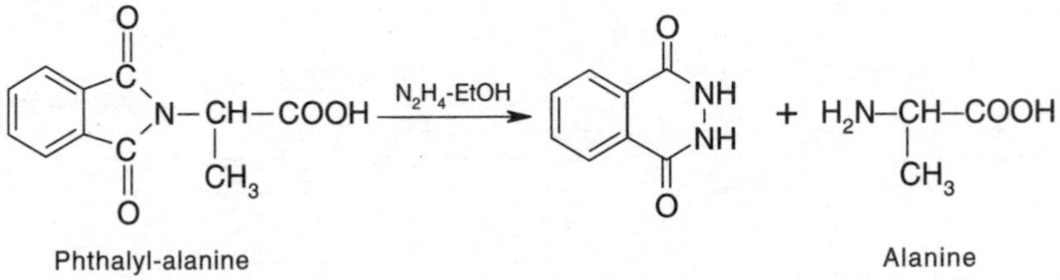

Phthalyl-alanine   Alanine

Tosyl (OTs) and dithiasuccinoyl (Dts) groups are some of the other N-protecting groups.

2.4.2 Protection of Carboxyl Group

The carboxyl group of amino acids is generally protected as their methyl, ethyl or benzyl esters, which can be prepared by the standard methods of ester formation. A solution of amino acid in alcoholic HCl is kept overnight to obtain crystalline ester hydrochloride. Alternatively, the carboxyl group of the amino acid is first converted into an acid chloride by using thionyl chloride and then an appropriate alcohol is added to get the ester.

O
||
$H_2NCHCOCH_2C_6H_5$ $C_6H_5CH_2OH/HCl$ CH_3OH/HCl O
| ||
CH_3 O $H_2NCHCOCH_3$

Alanine benzylester ||

                  $H_2NCHCOH$ CH_3
                    | Alanine methylester

                    CH_3 CH_3OH
                    Alanine   O
                                           ||
             $SOCl_2$ $H_2NCHCCl$

                                      CH_3
                                  Alanyl chloride

Since the amino acid esters do not have dipolar character, these are easily soluble in aprotic solvents –a quality highly desirable in peptide synthesis, as peptide bond formation in protic solvents is prone to racemisation.

Since it is relatively easy to hydrolyse esters than the amide linkages, these protecting groups are easily deprotected by mild hydrolysis with aqueous NaOH. The benzyl esters can also be cleaved by hydrogenolysis using palladium-charcoal catalyst or with the help of strong anhydrous acids like, HBr in acetic acid.

O
||
$H_2NCHCOCH_3$ (i) NaOH
| (ii) H_3O^+
CH_3

                                       O
                                       ||
                                   $H_2NCHCOH$
                                     |
                                     CH_3

O
||
$H_2NCHCOCH_2C_6H_5$ $H_2/Pd-C$
|
CH_3

2.4.3 Protection of Side Chains

As mentioned earlier, functional groups in the side chains of the amino acids can interfere in the process of peptide bond formation and need to be suitably protected. The side chain protecting groups are usually based on the benzyl (Bzl-) or the tertiary butyl (t-Bu-) group. Amino acids with alcohol or carboxyl group in the side chain are protected either as Bzl or t-Bu ethers or esters. A combination of Boc for the N-amino group and benzyl derivatives for the side chain protecting group has been used in most syntheses. For other types of functional groups, e.g. the thiol group of Cys, the imidazole group of His or guanidino group of Arg, certain specific protecting groups are required. It is difficult to generalise the protecting groups for the side chains because the choice of the protecting group depends to a great extent on the nature of protecting group chosen for the N-amino group and the coupling procedure. For every peptide to be synthesised, a strategy has to be worked out depending on the nature of the amino acids making the peptide.

2.4.4 Coupling Methods

Once the two amino acids are suitably protected the next step involves the generation of peptide bond between them. This step is called **coupling**. There are a number of methods used for this purpose; some of these are discussed here.

(i) Acid Chloride Method

This is by far the simplest approach. In this method the N-protected amino acid is activated by converting into its acid chloride, which is then made to react with an amino acid ester under Schotten-Baumen conditions to give the protected peptide. The desired peptide with free amino and carboxyl groups is obtained thereafter with mild hydrolytic reactions.

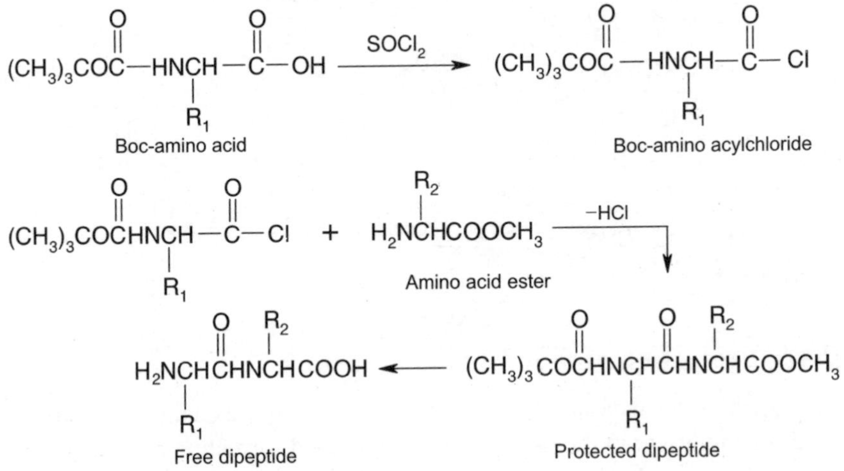

$$(CH_3)_3COC-HNCH-C-OH \xrightarrow{SOCl_2} (CH_3)_3COC-HNCH-C-Cl$$

Boc-amino acid   Boc-amino acylchloride

$$(CH_3)_3COCHNCH-C-Cl \ + \ H_2NCHCOOCH_3 \xrightarrow{-HCl}$$

Amino acid ester

$$H_2NCHCHNCHCOOH \longleftarrow (CH_3)_3COCHNCHCHNCHCOOCH_3$$

Free dipeptide   Protected dipeptide

(ii) Carbodiimide Method

This method was introduced by Sheehan and Hess in 1955. It involves the reaction of a N-protected amino acid with an amino acid ester with the help of an equimolar amount of dicyclohexylcarbodiimide (DCC), the coupling agent, in presence of an inert solvent like methylene chloride or THF, etc. Essentially, this coupling reagent promotes dehydration between the free carboxyl group of the N-protected amino acid and the free amino group of the C-protected amino acid, to give the protected peptide and a highly insoluble by-product, N, N´-dicyclohexylurea, which separates as a solid and can be filtered out.

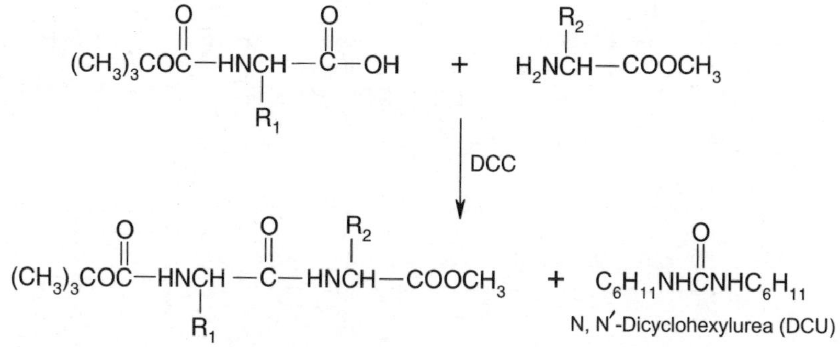

DCC can be prepared from carbon disulphide and cyclohexylamine as follows:

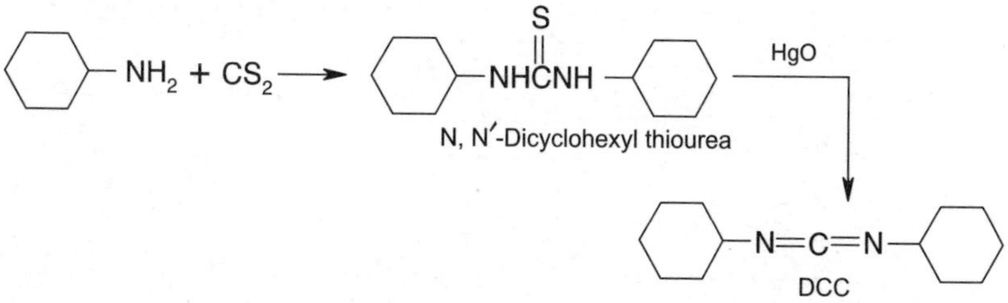

In the coupling reaction carbodiimide acts by activating the free carboxyl group of the N-protected amino acid by forming an O-acylisourea derivative as per the following reaction.

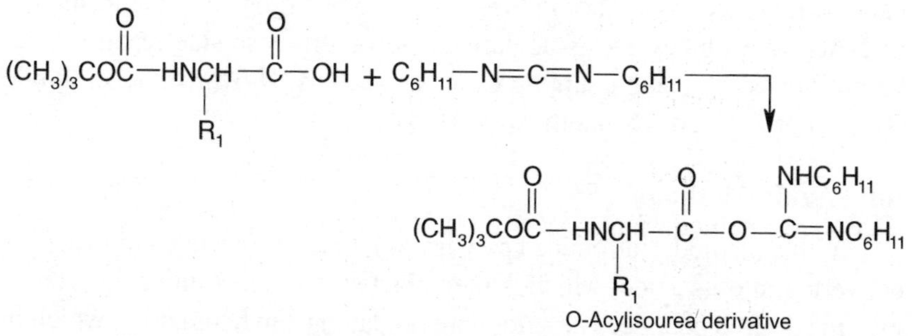

O-Acylisourea derivative

The activated acyl group of the first amino acid is then transferred to the amino group of the second amino acid to give the peptide and N, N´-dicyclohexylurea. The mechanism of the nucleophilic substitution reaction involved can be represented as:

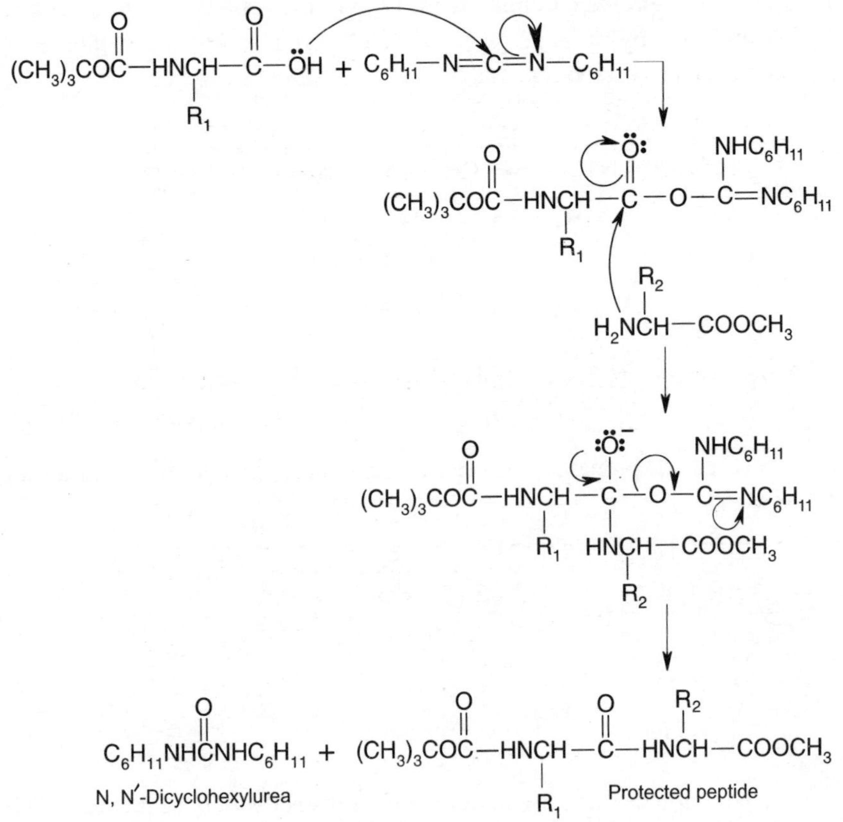

This DCC method of coupling is hampered by side reactions, which can result in racemisation or in the presence of a strong base, the formation of 5(4H)-oxazolones and N-acylureas. However, these side reactions can be minimised by using coupling catalysts such as, N-hydroxysuccinimide (HOSu) or 1-hydroxybenzotriazole (HOBt).

(iii) Azide Method

In this method, (introduced by Curtius), the N-protected amino acid is converted into its azide, which is then reacted with an amino acid ester to give the desired peptide. The azides are prepared from hydrazides, which in turn are prepared by hydrazinolysis of protected amino acid ester.

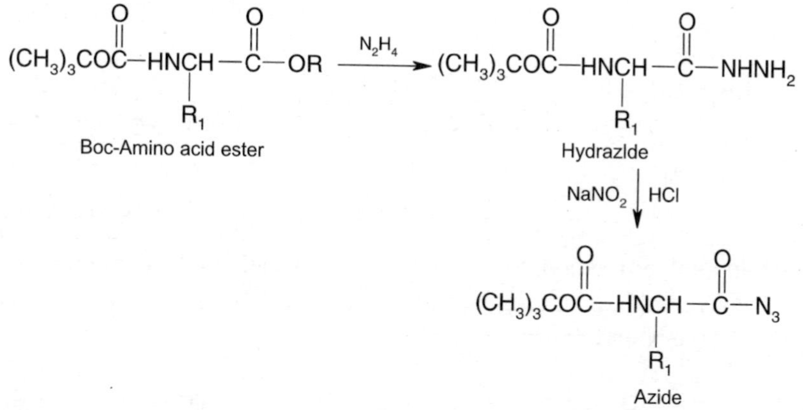

$$(CH_3)_3COC-HNCH-\overset{O}{\overset{\|}{C}}-OR \xrightarrow{N_2H_4} (CH_3)_3COC-HNCH-\overset{O}{\overset{\|}{C}}-NHNH_2$$

Boc-Amino acid ester   Hydrazide

$$NaNO_2 \Big| HCl$$

$$(CH_3)_3COC-HNCH-\overset{O}{\overset{\|}{C}}-N_3$$

Azide

The azide so obtained is immediately coupled with the ester of second amino acid (aminolysis).

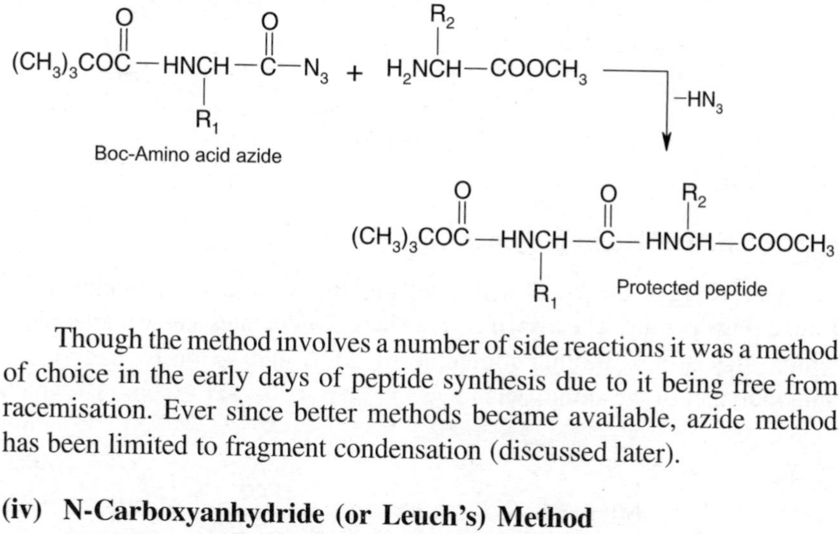

$$(CH_3)_3COC-HNCH-\overset{O}{\overset{\|}{C}}-N_3 + H_2NCH-COOCH_3$$

Boc-Amino acid azide

$$-HN_3$$

$$(CH_3)_3COC-HNCH-\overset{O}{\overset{\|}{C}}-HNCH-COOCH_3$$

Protected peptide

Though the method involves a number of side reactions it was a method of choice in the early days of peptide synthesis due to it being free from racemisation. Ever since better methods became available, azide method has been limited to fragment condensation (discussed later).

(iv) N-Carboxyanhydride (or Leuch's) Method

In this method (more appropriate for making homopolymers of amino acids) N-carboxyanhydride (NCA) of an amino acid is prepared and polymerised by heating with a catalyst in an organic acid. However, this method can also be used to prepare peptides by sequentially adding amino acids.

Preparation of NCA

The N-carboxyanhydride can be prepared in a number of ways.

(a) Reaction of an amino acid with phosgene gives N-carbonyl chloride intermediate which cyclises to give N- carboxyanhydride (NCA). The reaction for the preparation of glycine NCA can be written as follows.

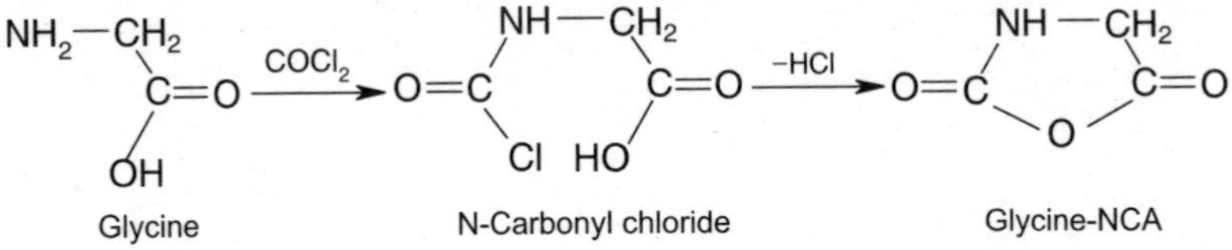

Glycine   N-Carbonyl chloride   Glycine-NCA

(b) In another method, the N-benzyloxy derivative of the amino acid is converted into an acid chloride, which on heating in vacuum cyclises to give N-carboxyanhydride.

N-benzyloxyglycine

heat in vacuum

Glycine-NCA

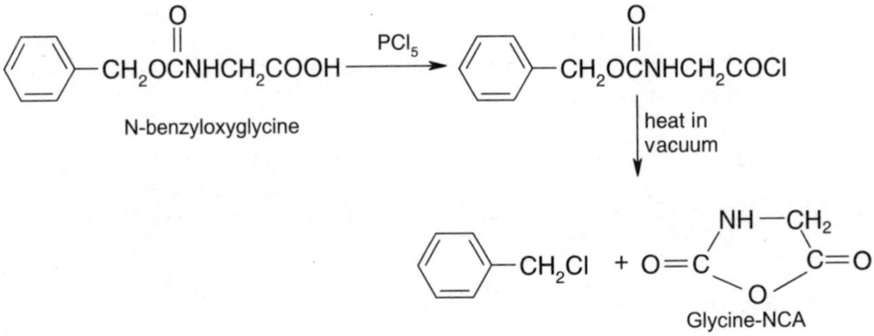

Preparation of Homopolymer

The NCAs on treatment with small amounts of initiator (a nucleophile) lead to ring opening. The loss of carbon dioxide generates a new nucleophile which may attack another molecule of NCA and so on to generate a homopolymer of the amino acid.

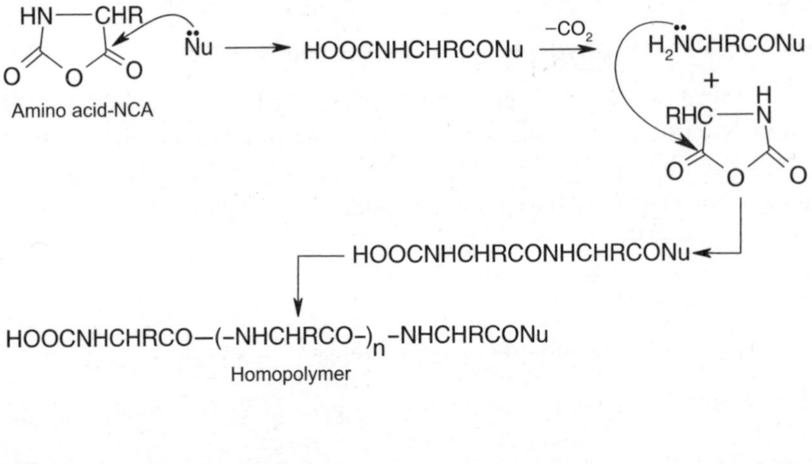

Amino acid-NCA

$$HOOCNHCHRCONu \xrightarrow{-CO_2} H_2NCHRCONu$$

$$HOOCNHCHRCONHCHRCONu$$

$$HOOCNHCHRCO-(-NHCHRCO-)_n-NHCHRCONu$$

Homopolymer

Preparation of Peptide

In order to make a peptide the amino acid NCA is combined with an amino acid under carefully controlled alkaline conditions. The product so obtained is then acidified to get the dipeptide.

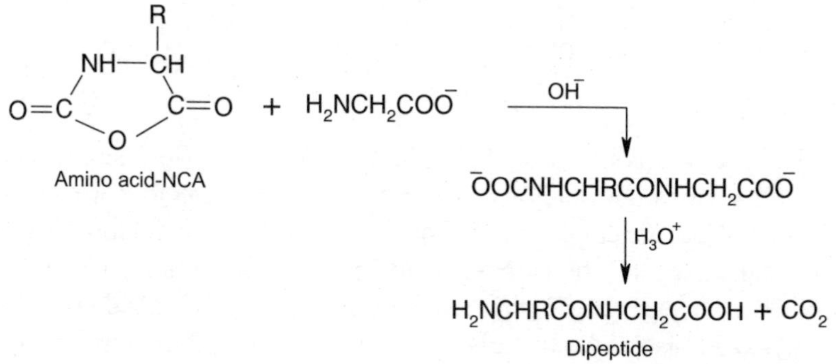

$$\underset{\text{Amino acid-NCA}}{\underset{\displaystyle O=C\diagdown_{\displaystyle O}\diagup C=O}{\overset{\displaystyle NH-\overset{\displaystyle R}{\underset{|}{C}H}}{}}} \quad + \quad H_2NCH_2COO^- \quad \xrightarrow{\;\; OH^-\;\;}$$

$$^-OOCNHCHRCONHCH_2COO^-$$

$$\Big\downarrow H_3O^+$$

$$H_2NCHRCONHCH_2COOH + CO_2$$
$$\text{Dipeptide}$$

The peptide chain can be extended by reacting the dipeptide with another amino acid NCA. The peptide chain grows towards the N-terminal direction. The peptides obtained by this method are of high optical purity. However, problem may arise, the carbamic acid produced in the process of aminolysis is unstable and can undergo decarboxylation to give free amino acid. This free acid can react with the NCA to give undesired product.

(v) Mixed Carbonic Anhydride Method

In this method a mixed carbonic anhydride is prepared from an alkyl chloroformate like ethylchloroformate or isobutylchloroformate which then combines with a free amino group of an amino acid or a peptide unit. This method is used in cases where separation of the by-product, dicyclohexylurea (in DCC method) proves to be difficult. The first step of activating the carboxyl group of a N-protected amino acid takes place in an organic solvent in the presence of a tertiary base like, triethylamine.

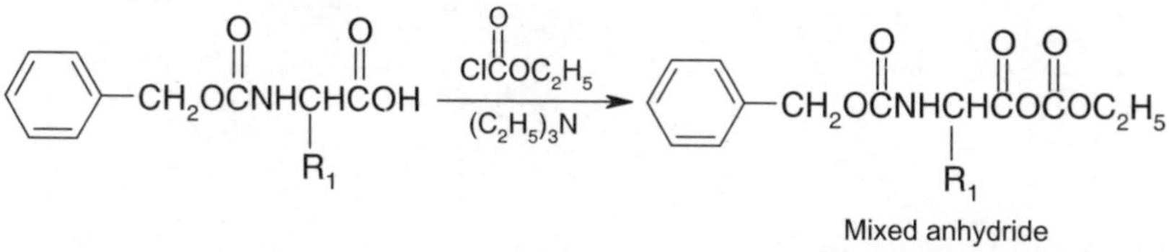

Mixed anhydride

In the second step involving peptide formation, the anhydride is usually added at a 14-fold excess over the amino component.

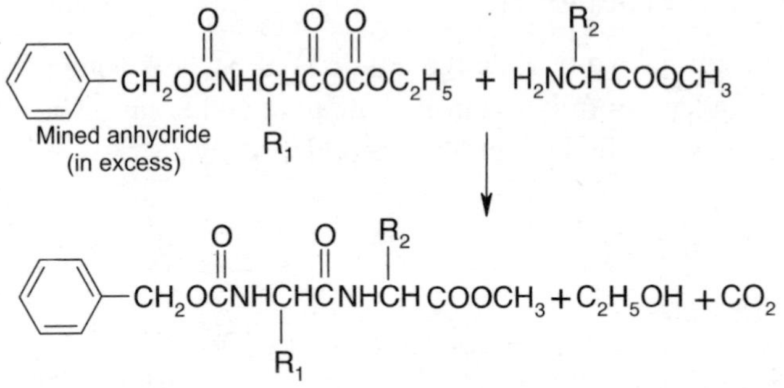

$$
\underset{\substack{\text{Mined anhydride}\\\text{(in excess)}}}{\boxed{\bigcirc}-CH_2OCNHCHCOCOC_2H_5} \;+\; H_2NCHCOOCH_3
$$

with O (three carbonyl groups), R_1 on the first chain and R_2 on the amino component.

$$
\boxed{\bigcirc}-CH_2OCNHCHCNHCH\,COOCH_3 + C_2H_5OH + CO_2
$$

with R_1 and R_2 substituents.

This method is noted for being highly effective at low temperatures ($\sim -15°C$), giving high yields and pure products when protecting groups such as Cbz or *t*-Boc, are employed. Although isobutyl and ethylchloroformates are typically used to form carbonic anhydrides, other coupling reagents are also known. For example, N-ethyloxycarbonyl-2-ethoxy-1, 2-dihydroquinoline (EEDQ) and N-isobutyloxy-carbonyl-2-isobutyloxy-1, 2-dihydroquinoline (IIDQ).

The mixed carbonic anhydride procedure using chloroformate invokes separate steps for activation and aminolysis. However, if we use EEDQ to prepare the mixed carbonic anhydride, the reaction can be performed in the presence of the amino component which combines with the anhydride as soon as it is formed. This method can be used as a direct coupling method by which the possibility of the side reaction is also minimised.

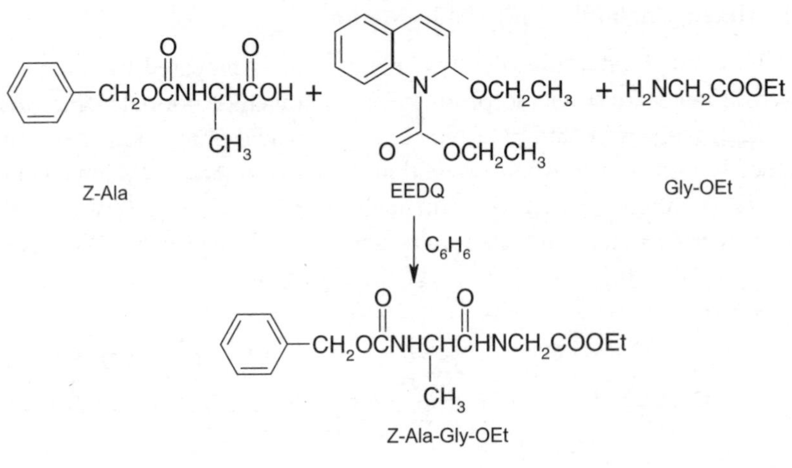

$$
\underset{\text{Z-Ala}}{\boxed{\bigcirc}-CH_2OCNHCHCOH}_{\;CH_3} + \underset{\text{EEDQ}}{\text{(dihydroquinoline-OCH}_2CH_3)} + \underset{\text{Gly-OEt}}{H_2NCH_2COOEt}
$$

$$
\downarrow C_6H_6
$$

$$
\underset{\text{Z-Ala-Gly-OEt}}{\boxed{\bigcirc}-CH_2OCNHCHCHNCH_2COOEt}_{\;CH_3}
$$

(vi) Azolide Method

In this method, the N-protected amino acid is converted into its imidazolide by reacting the amino acid with an equimolar quantity of

N, N´-carbonyldiimidazole in THF at room temperature.

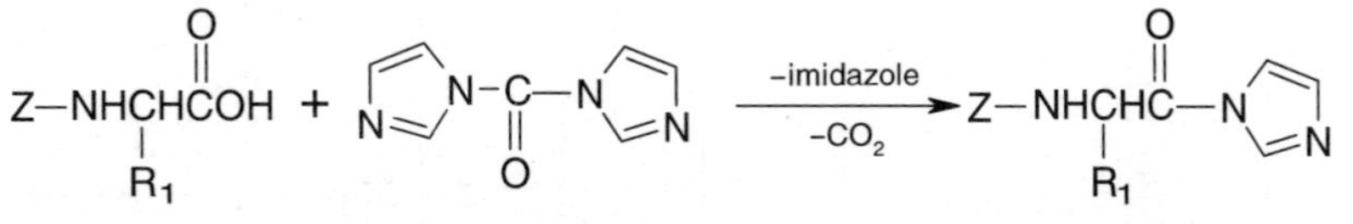

N,N'carbonyldiimidazole   Imidazolide

The energy rich amino acid imidazolide is then reacted with an amino acid ester to give the desired peptide.

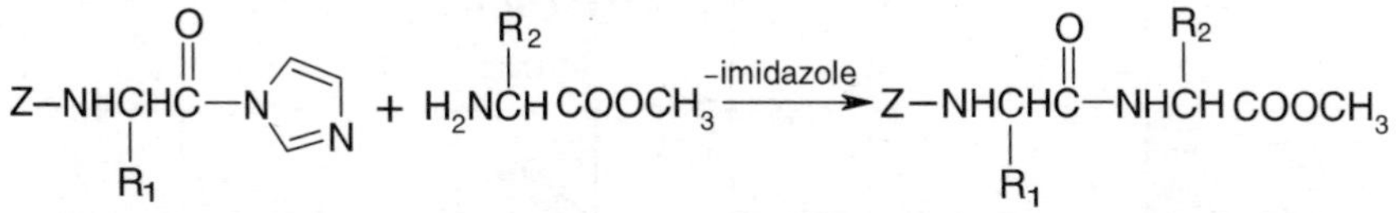

The "classical" methods for the synthesis of peptides in solution (discussed above) are quite time consuming and skill intensive, due to the unpredictable solubility characteristics of intermediates involved, yields and possible side reactions. The synthesis of a peptide of a reasonable size by this approach requires many steps. The requirement of purifying the product obtained at each step coupled with the material losses during handling make solution phase peptide synthesis a highly skill intensive endeavour. The synthesis of oxytocin–a peptide hormone involved in the utrine contraction during labour and lactation, by Vigneaud and Bodansky as per the scheme given in Fig.2.2, outlines a typical synthetic strategy.

To circumvent the problems of solution phase synthesis, R. Bruce Merrifield devised a clever strategy called solid phase peptide synthesis.

2.5 SOLID PHASE PEPTIDE SYNTHESIS

The **solid phase peptide synthesis** (SPPS) also known as the **Merrifield synthesis** after its inventor uses an insoluble polymeric resin as a solid support for synthesising the peptide. Polystyrene cross linked with about 2% divinyl benzene is chloromethylated to make the resin. The C-terminal of N-protected amino acid of the peptide to be synthesised is then linked to the polymer by nucleophilic displacement. The resin swells in commonly used solvents for peptide synthesis and forms a three dimensional matrix in which the reagents can penetrate freely.

Solid phase peptide synthesis consists of three distinct sets of operations as given below and detailed after that :

- Chain assembly on the resin
- Cleavage and deprotection of the peptide chain
- Purification and characterisation of the target peptide

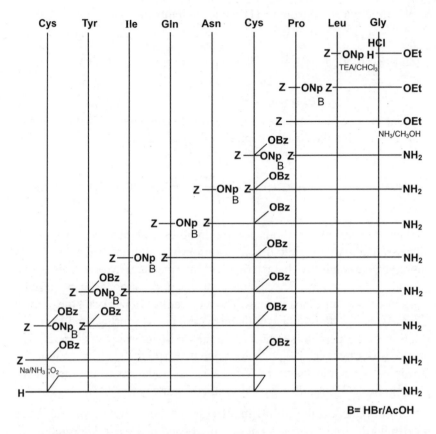

Fig. 2.2 *Scheme outlining the strategy of Vigneaud and Bodansky for synthesising oxytocin. The peptide is synthesised sequentially from the C-terminal. The protecting and activating groups and deprotecting reagents are suitably indicated. The intramolecular disulphide linkage between cysteine side chains is created in the last step.*

Chain Assembly on the Resin

The chain assembly involves covalently attaching the N-protected, C-terminal amino acid to the polymeric solid matrix which is in the form of small beads. This is called **anchoring**. The N-protection of the linked amino acid is removed by mild acid and neutralised to give a free amino group. The next amino acid, with a protected α-amino group, is activated and reacted with the resin bound amino acid to yield an amino protected dipeptide on the resin. This process of removing the N- protection followed by neutralisation and coupling of the activated amino acid derivatives is continued stepwise, extending the peptide chain towards N-terminal, to obtain the desired peptide on the resin. The polymeric resin keeps the growing peptide insoluble and also provides protection to the C-terminal. The separation and purification at different steps is accomplished simply by filtering and washing the beads with appropriate solvents. It removes the soluble by-products and the excess reagent while the peptide remains anchored to the resin. The general strategy of solid phase peptide synthesis is given in Fig. 2.3.

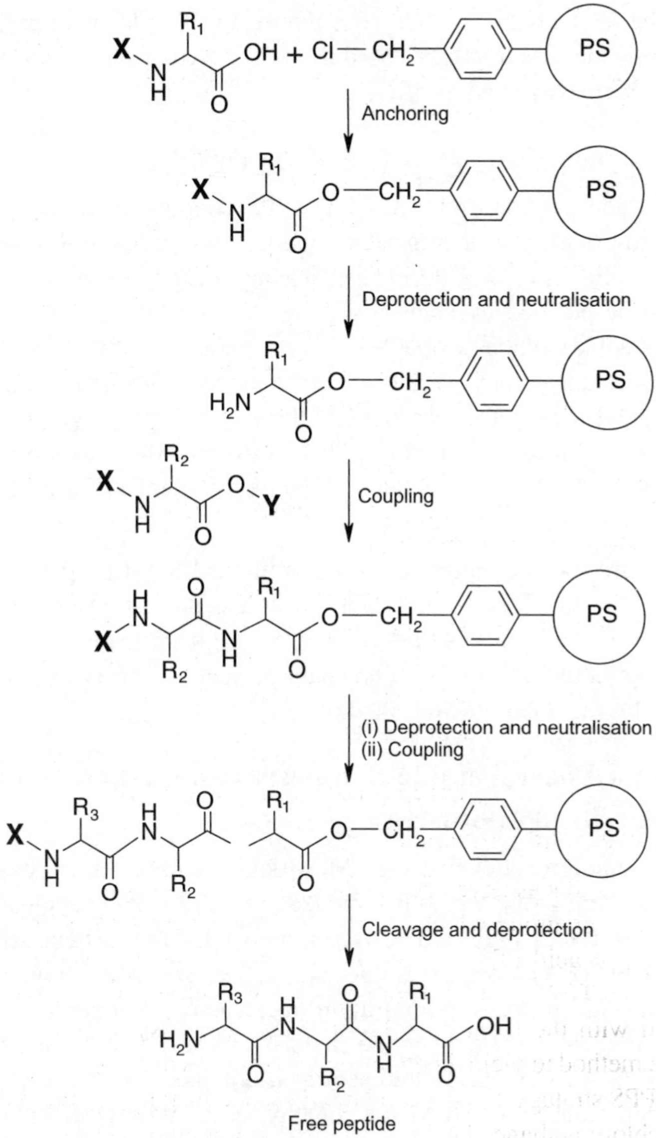

Fig. 2.3 *General strategy for solid phase peptide synthesis. Encircled PS represents polymeric solid matrix while X and Y refer to protecting and activating groups respectively.*

Cleavage and Deprotection of the Peptide Chain

The cleavage and deprotection of the synthesised peptide chain can be done simultaneously or the two can proceed sequentially. Acidolysis is the most common method used to cleave the assembled peptide from the resin. Usually this procedure simultaneously deprotects the side chain protections also. In the sequential approach, after removing the side chain protections (if any) the peptide is finally released from the polymer support by a simple benzyl ester cleavage and purified.

Purification and Characterisation of the Peptide

The crude peptide obtained from SPPS contains many by-products resulting from deletion or truncated peptides (Sec. 2.5.3) and chemically modified peptides obtained during the cleavage and deprotection processes. These can be purified by using a suitable combination of wide range of chromatographic methods available like, size-exclusion chromatography, ion-exchange chromatography, partition chromatography, high performance liquid chromatography, preparative paper and thin layer chromatography etc. Of these, reverse-phase (RP) HPLC is the most versatile and most widely used method. The purified peptides are then characterised with the help of available spectroscopic techniques.

There are two strategies commonly followed for solid phase peptide synthesis. One involves an acid labile α-amino protecting group (t-Boc) while the other uses a base labile protecting group (Fmoc). These use different protecting groups for side chain protection and correspondingly different cleavage/deprotection methods.

2.5.1 Solid Phase Peptide Synthesis Using *t*-Boc Protection (Merrifield Approach)

This method was developed by Merrifield and uses t-Boc group as the amino protecting group which is removed at every cycle by acidosis using trifluoroacetic acid (TFA). The free amino terminal is neutralised by triethylamine (TEA). The next amino acid, with a protected α-amino group, is coupled with the resin bound amino acid using symmetric or mixed anhydride method to yield dipeptide on the resin. As discussed earlier in the general SPPS strategy, the desired peptide is obtained by repitition of these steps. Dichloromethane (DCM) or dimethylformamide (DMF) is used as the primary solvent for deprotection, coupling and washing procedures. Excess reactants and by-products are removed by simple filtration and washing. Since TFA is used repeatedly for the deprotection, the side chain protection (if required) is done with the groups that can withstand the TFA cleavage at every cycle. Benzyl or cHex based protecting groups are normally used in the side chain protection in such cases. This approach of using Boc group for the N-protection and benzyl for the side chain is sometimes referred to as Boc/Bzl strategy of SPPS. The cleavage of the peptide from the resin is done with the help of liquid HF. This removes the N-terminal *t*-Boc group also.

There are two problems with this approach. Firstly, in the course of repeated Boc cleavage (while synthesising long peptides) there is a possibility of the side chain deprotection. This can be minimised by using a more acid resistant benzyl derivative. The second problem is the danger of cleavage of the benzyl linkage of the peptide with the resin. This problem can be minimised by using a polymer support (PAM resin) to which the first amino acid is attached through a more acid resistant benzyl ester linkage.

The Boc/Bzl- strategy of SPPS using PAM resin can be illustrated for the preparation of a tripeptide Tyr-Gly-Ala as detailed in Fig.2.4.

Though HF is commonly used for cleaving the peptide from the resin other alternatives like, trifluoromethanesulphonic acid (TFMSA) and trimethylsilyltrifluoromethanesulphonate (TMSOTF) are also available.

2.5.2 Solid Phase Peptide Synthesis Using Fmoc Protection (Sheppard's Approach)

This method was developed by Sheppard and his associates and uses Fmoc group as the amino protecting group which is removed at every cycle by organic bases like piperidine solution in DMF. In Fmoc-strategy active ester method is the most widely used coupling method. The active esters of the Fmoc derivatives are preformed or are generated *in-situ* during the coupling step. Initially, the *p*-nitrophenyl and N-hydroxysuccinimide activated (ONSu) esters were routinely used however, nowadays the pentafluorophenyl (OPfp) and the 3-hydroxy-2,3-dihydro-4-oxo-benzo-triazone (ODhbt) esters are the most commonly used esters. The side chain protection (if required) is done with the help of acid labile groups; *t*-But being the group of choice. Such a scheme of protection is referred to as orthogonal protection–a highly desirable situation in SPPS.

In Sheppard's approach, a polyamide resin is used in contrast to the polystyrene based resin of Merrifield. The cross linked polyacrylamide based copolymer contains sarcosine methyl ester side chains which are coupled with Fmoc norleucine (leucine residue with one methyl group less in the side chain). This is followed by a linker or a resin handle to which the first

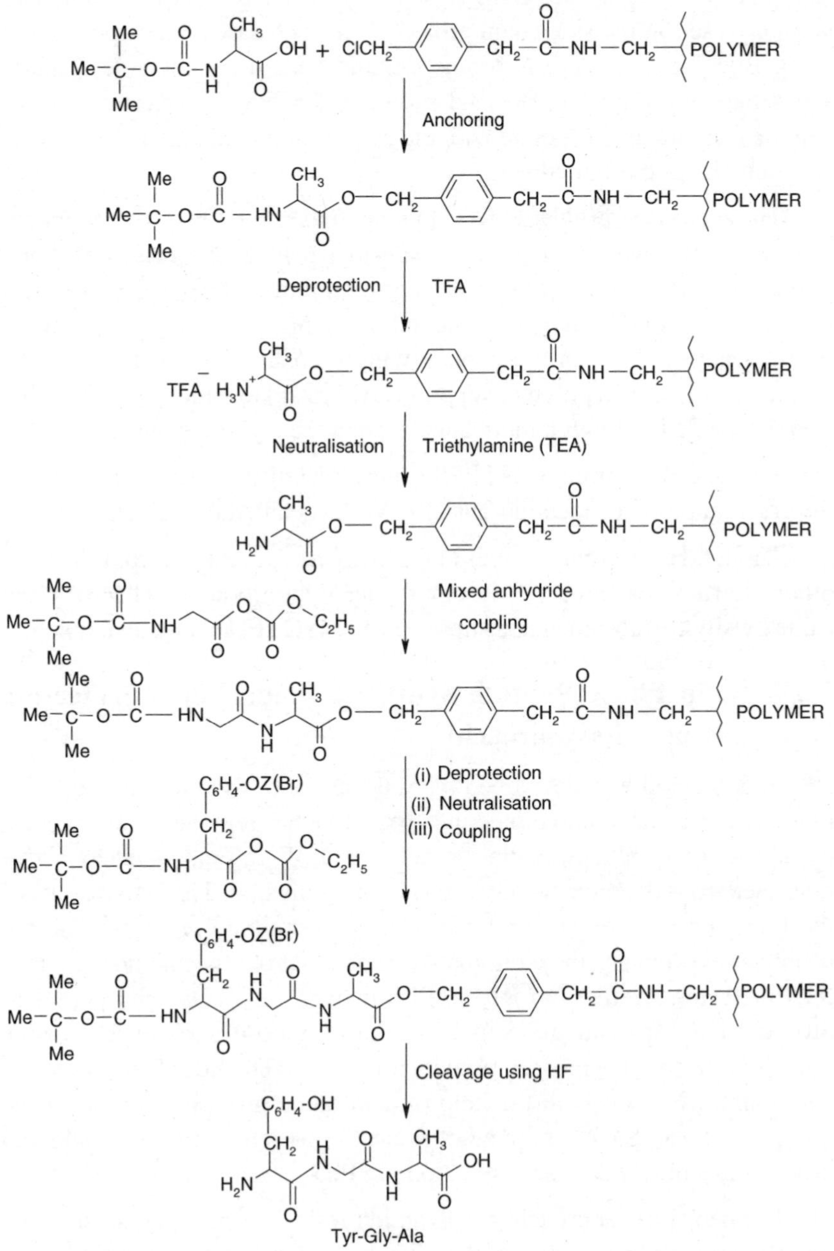

Fig. 2.4 *Scheme for synthesising the tripeptide, Tyr-Gly-Ala using Boc/Bzl- strategy of Merrifield. Here Z(Br)-stands for bromobenzyloxycarbonyl group.*

amino acid (C-terminal) of the peptide to be synthesised is attached. The norleucine residue provides internal reference of the amino acid while the resin and the handle make the polymeric support to be chemically similar to the peptide being synthesised. This helps in better access of the reagents in the polymer matrix and better solvation of the matrix. A highly solvated polymer matrix leads to lesser aggregation of the peptide being synthesised- a drawback in the Merrifield approach. The Fmoc /-OBu^t strategy of SPPS can be illustrated for the preparation of a tripeptide, Ala-Ala-Tyr as given in Fig.2.5.

Due to the speed and simplicity of the repeated steps, the major portion of the solid phase procedure can be automated and for the reasons of efficiency and convenience, the solid phase peptide synthesis (SPPS) has proven to be the method of choice for producing peptides and small proteins of specific sequences. However there are certain limitations too.

2.5.3 Limitations of Solid Phase Peptide Synthesis

The first and foremost requirement of a successful SPPS synthesis is that the deprotection and the coupling steps must proceed to almost 100% completion. The uncoupled amino groups get acylated in the subsequent cycle resulting in the formation of what are called as **deletion peptides**. Similarly, if the deprotection is not complete, the left over groups may get removed in subsequent cycle and couple with the corresponding amino acid, again leading to a deletion peptide. Further, there is a possibility that the free amino group does not get coupled in the given or subsequent cycles and lead to the formation of a **truncated peptide**. The incomplete deprotection or coupling or both have cumulating effect over a large number of cycles involved in the synthesis. The resulting product is a mixture of closely related peptides differing only slightly from the targeted peptide and pose a real problem of separation. Sometimes the isolation of the desired peptide from the microheterogenous mixture becomes almost impossible.

In addition to the generation of a mixture of peptides another consequence of possible inefficiency of the coupling steps is the low yield. If the overall efficiency of adding each successive amino acid is 90%, the yield of a 20 residue peptide would be a just 12% which would reduce to 0.003% for a 100 residue peptide. For a 100 residue peptide a 99% efficiency (highly ambitious) of each step would give an overall yield of only 37%. This problem of inefficiency of stepwise synthesis can be tackled by resorting to what is called as the **fragment condensation**.

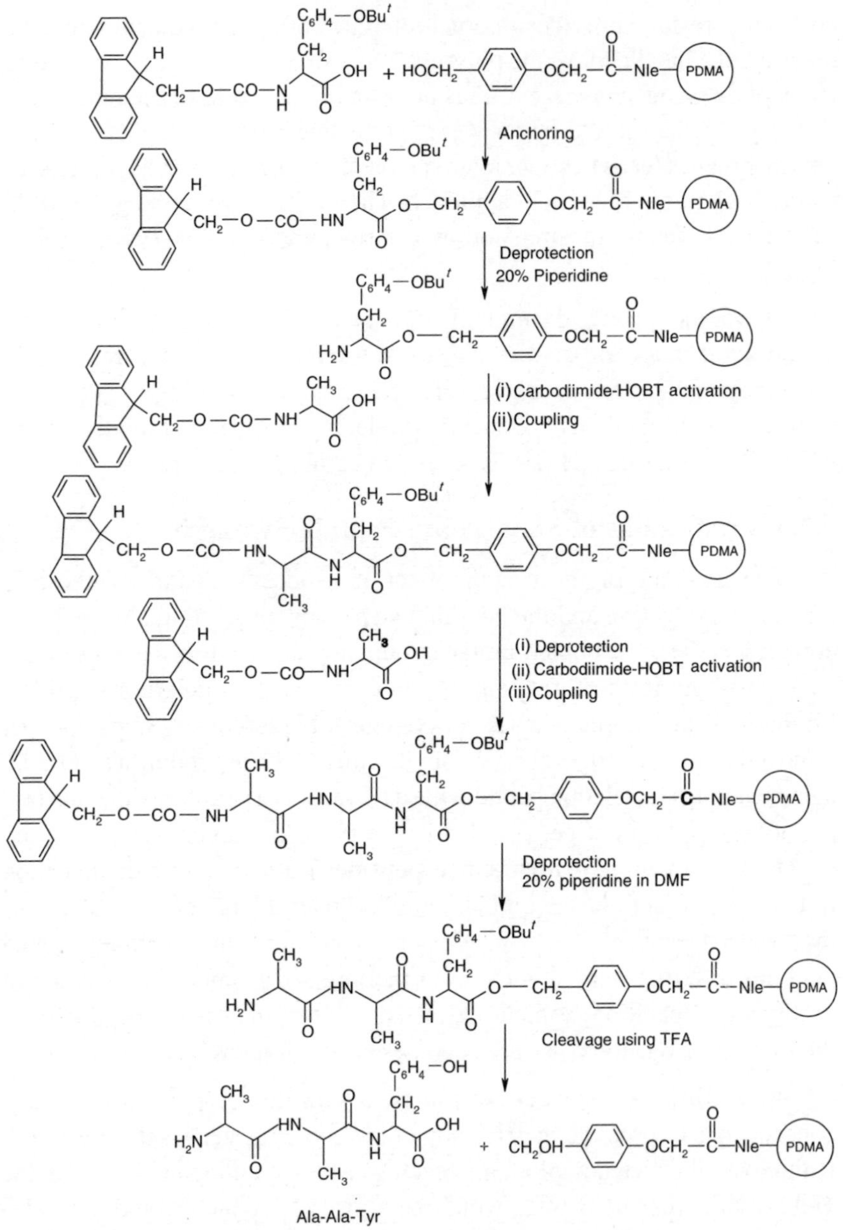

Fig. 2.5 *Scheme for synthesising the tripeptide, Ala-Ala-Tyr using Sheppard's strategy. OBuᵗ represents t-butyl ether derivative of the tyrosyl side chain. Nle refers to norleucine –an internal reference in amino acid analysis of the systhesised peptide and PDMA is the resin containing polydimethylacrylamide.*

In this approach two or more moderately sized peptides are synthesised by stepwise procedure and are joined together by selective peptide bond formation. The advantage of this method lies in the fact that the overall yield of small fragments is good and these can be purified before condensation. Further, the product of fragment condensation can be purified easily because the resulting mixture contains peptides of different sizes. For example, in the condensation of an octapeptide with a dodecapeptide (12 residue peptide) to give a 20 residue peptide, the crude peptide mixture would contain peptides of 8,12 and 20 residues besides the corresponding by-products.

Another area of concern in the solid phase peptide synthesis is the possibility of racemisation and other side reactions. The recent developments in the area provide a wide choice of resins, protecting groups, coupling methods, deprotection and cleavage reagents and strategies. A right mix of these can be used to mininise the problems of undesired side reactions.

2.6 SOME BIOLOGICALLY IMPORTANT PEPTIDES

Natural peptides are of varying complexity and provide a great deal of structural diversity. The widely distributed antioxidant tripeptide - glutathione is interesting for the fact that in this molecule the side chain carboxyl function of the N-terminal glutamic acid is used for the peptide bond in place of usual carboxyl group. In some of the peptides two or more cysteine residues present at different positions in a peptide chain, are often joined by disulfide bonds e.g. oxytocin, and in case of insulin, two separate peptide chains (A and B chains) are held together by such links. Further, in some of the peptides the C-terminal unit is in the form of an amide e.g., oxytocin. Besides structural diversity, the peptides also have a wide ranging functional diversity. We will discuss some of the peptides with important biological functions and interesting structural features.

2.6.1 Oxytocin

Oxytocin is a peptide hormone, containing nine amino acids, that is synthesised in hypothalamic neurons and transported to the pituitary gland for secretion into the blood. It is also secreted within the brain and from a few other tissues, including ovaries and testes. Structurally, the peptide contains a disulphide linkage between cysteine residues at position 1 and 6 and the C-terminal carboxyl is in the form of an amide.

Cys–Tyr–Ile–Gln–Asn–Cys–Pro–Arg–Gly–NH$_2$

The structure of the peptide was established independently by Du Vigneaud and Tuppy in 1953 through two independent approaches involving performic acid oxidation (to break disulphide linkages), controlled hydrolysis followed by separation and sequence analysis of the fragments obtained. The structure was later confirmed by synthesis in 1954 by Du Vigneaud.

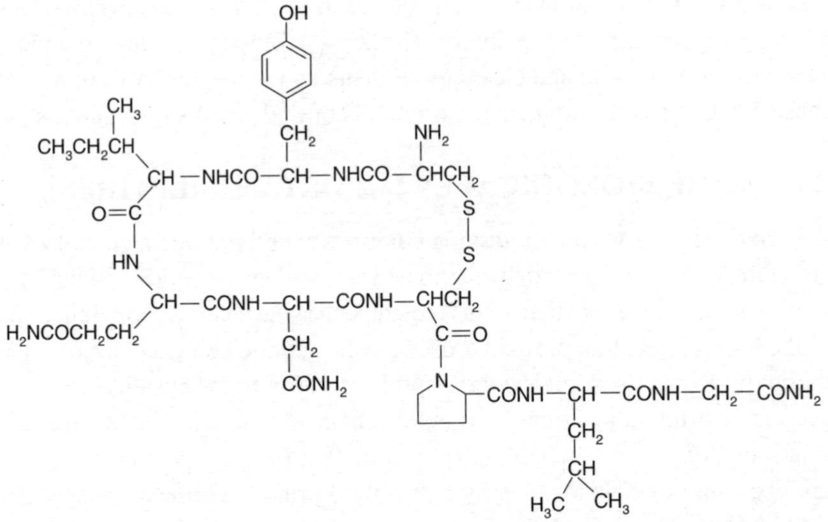

Oxytocin has a few well-defined activities related to birth and lactation. These are, stimulation of milk ejection, and uterine smooth muscle contraction at the time of parturition and invoking of maternal behaviour towards the child soon after giving birth. In addition to these, it has many subtle but profound influences like it is concerned with the feeling of satiation (fullness) when the food is absorbed and it has also been shown to be involved in facilitating sperm transport within the male reproductive system. The three-dimensional structure of oxytocin is given in Fig. 2.6.

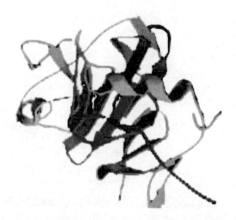

Fig. 2.6 *The three-dimensional structure of the peptide oxytocin. The meaning of the curves and arrows are discussed in chapter 3 (P-157).*

The action of oxytocin involves the N-terminal amino group, the tyrosyl hydroxyl group, the disulphide linkage and the amide functional groups in the side chains of asparagine, glutamine and the C-terminal glycinamide. The small size of oxytocin has permitted the synthesis of numerous analogs in which the functionally important parts of the peptide are either removed or modified and have shown properties similar to or better than oxytocin. In cases where uterine contractions are not sufficient to complete delivery, physicians and veterinarians sometimes administer oxytocin to stimulate uterine contractions so as to facilitate delivery of the child.

2.6.2 Glutathione

Glutathione (γ-glutamylcysteinylglycine) is a tripeptide found in nearly all cells of plants, animals and microorganisms. The presence of glutathione in the body is required to maintain the normal functioning of the immune system. It is known to play a critical role in the multiplication of lymphocytes (the cells that mediate specific immunity) which occurs in the development of an effective immune response. Glutathione is also used to prevent oxidative stress in most cells and helps to trap free radicals that can damage DNA and RNA. There is a direct correlation with the speed of aging and the reduction of glutathione concentrations in intracellular fluids. As individuals grow older glutathione levels drop and the ability to detoxify free radicals decreases.

Structurally, glutathione is an unusual peptide in the sense that the N-terminal glutamyl residue forms a peptide bond through its γ-carboxyl group (in the side chain) instead of the usual carboxyl group.

$$H_3\overset{+}{N}-CH-CO\overset{-}{O}$$
$$CH_2$$
$$CH_2$$
$$CONH-CH-CONH-CH_2-COOH$$
$$CH_2SH$$

Glutathione

This atypical (unusual) peptide bond formation probably is to escape degradation in the body. The synthesis of glutathione in the cell involves two steps. In the first step glutamic acid condenses with cysteine in a reaction catalysed by γ-*glutamylcysteinesynthase*.

Glutamate + Cysteine + ATP ⟶ γ-glutamylcysteine + ADP + P_i

In the second step this dipeptide condenses with glycine to give glutathione in the presence of enzyme, *glutathionesynthase*.

γ-glutamylcysteine + Glycine + ATP ⟶ γ-glutamylcysteinylglycine + ADP + P_i

Glutathione performs a number of important functions in the body, the most significant being as an antioxidant and as a molecule responsible for forming disulphide linkages in the protein and thereby in protein folding. Glutathione is also involved in the transport of amino acid and trace minerals across cell membranes and may be involved in the repair of damaged cells. Two important roles of glutathione are discussed here.

2.6.2.1 Role of Glutathione in Disulphide Bond Formation

In the course of peptide or protein synthesis in the cell, the polypeptide chain is synthesised as a linear polymer. The cysteine residues in the peptide have free -SH groups in their side chains and are placed far apart in their primary structure. The job of disulphide bond formation between these isolated cysteine residues is performed by this tripeptide. The resulting bridging between the cysteine residues is very important for the tertiary structure of the protein. The exact mechanism of this process is not fully known, however, the role of glutathione cannot be underestimated. Glutathione can exist both as a thiol (GSH) as well as a disulphide (GSSG). The oxidised glutathione, GSSG, acts on suitably oriented cysteine residues of the polypeptide chain and undergoes two sequential thiol- disulphide exchanges as outlined below to generate the disulphide bond in the protein.

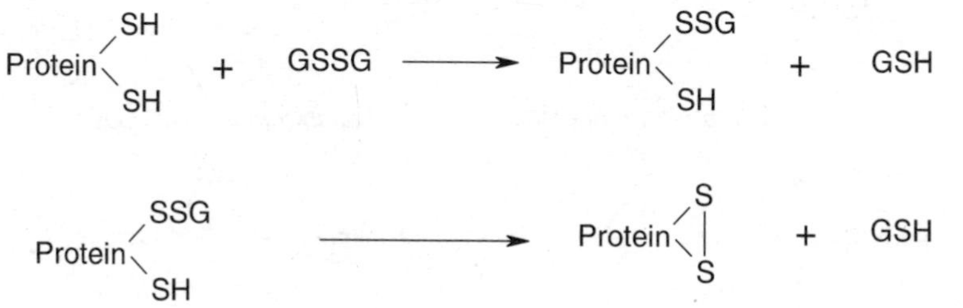

$$\text{Protein}\big\langle{\text{SH} \atop \text{SH}} \quad + \quad \text{GSSG} \quad \longrightarrow \quad \text{Protein}\big\langle{\text{SSG} \atop \text{SH}} \quad + \quad \text{GSH}$$

$$\text{Protein}\big\langle{\text{SSG} \atop \text{SH}} \quad \longrightarrow \quad \text{Protein}\big\langle{\text{S} \atop \text{S}}\big| \quad + \quad \text{GSH}$$

Fig. 2.7 *Disulphide bond formation in proteins: Role of Glutathione. GSSG and GSH are the oxidised and reduced forms of glutathione respectively.*

2.6.2.2 Role of Glutathione as an Antioxidant

A number of potentially harmful oxidising agents are generated in our body. These are the by-products of metabolism and also result from exposure to the radiations from the sun, X-rays, pollution, stress, physical exertion and due to various diseases. These include super oxides (O_2^-), hydrogen peroxide (H_2O_2), peroxy radicals ($ROO^{\cdot}$) and hydroxide radical ($OH^{\cdot}$). These free radicals are supposed to be involved in heart disease, stroke, Alzheimer's disease, Parkinson's disease, cancer, cataract and many other health problems. Glutathione acts as the first line of defence of the cell against such potentially harmful oxidising agents. The reduced glutathione reacts with these oxidising agents and gets oxidised in the process.

The accumulation of H_2O_2, a metabolic by-product in erythrocytes can affect its life span by increasing the rate of oxidation of haemoglobin to methemoglobin. The reduced glutathione causes the destruction of this H_2O_2 with the help of an enzyme called *glutathione peroxidase* which uses selenium as a prosthetic group. This is important in ensuring the proper functioning of haemoglobin, membrane lipids and other proteins.

$$H_2O_2 \;+\; 2GSH \xrightarrow{\;\text{glutathione peroxidase}\;} GSSG \;+\; 2H_2O$$

The oxidised glutathione (GSSG) obtained in the above process is reduced back with the help of reduced coenzyme NADPH, in a reaction catalysed by the enzyme *glutathione reductase*.

$$GSSH \;+\; NADPH \;+\; H^+ \xrightarrow{\;\text{glutathione reductase}\;} 2GSH \;+\; NADP^+$$

These reactions can be summarised as follows:

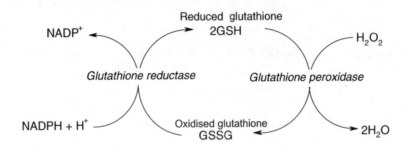

2.6.3 Insulin

Insulin is a peptide hormone, secreted by *islets of langerhans* (an endocrine gland) in the pancreas and is involved in controlling the blood glucose level. It was the first protein whose amino acid sequence was worked out almost half a century back by F. Sanger. It has been shown to be one of the smallest protein molecule (relative molecular mass of about 5.7-kD) having just 51 amino acid residues on two different chains of 21(A-chain) and 30 (B-chain) residues linked through disulphide bonds. The side chains of Cys^7 and Cys^{20} residues of the A-chain form the disulphide linkage with the Cys^7 and Cys^{19} residues of the B-chain. In addition, A-chain has an intra-chain S-S loop between Cys^6 and Cys^{11} residues.

2.6.3.1 Structure Determination of Insulin

The relative molecular mass of the purified insulin was found to be 5734 Da. The determination of N-terminal residue by DNP method gave glycine and phenylalanine as the outcome indicative of two chains. Further, since the amino acid analysis showed the presence of cysteine, the two chains were assumed to be linked through disulphide linkages.

The oxidation of insulin with performic acid gave two separate chains. These chains were separated and analysed using partial hydrolysis with acid and enzymes to get smaller fragment which were further analysed using standand methods of sequence determination. The fragments so obtained were separated and analysed for their sequence using standard methods (Sec 3.4.1.2). The sequences of peptide fragments, with overlapping regions were reassembled to determine the overall sequence. The position of the disulphide linkages was established by cleaving the purified insulin (without cleaving the disulphide linkages) again with the enzymes and repeating the sequence determination all over.

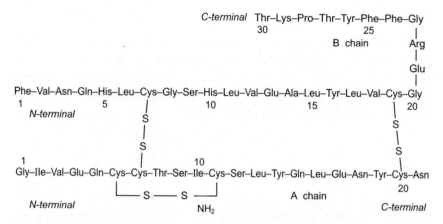

Fig. 2.8 *The structure of human insulin showing the inter (between Cys[7] of A-chain and Cys[7] of the B-chain and Cys[20] of A-chain and Cys[19] of the B-chain) and intra (between Cys[6] and Cys[11] of the A-chain) chain disulphide linkages.*

The amino acid sequence of insulin from a number of sources (sheep, horse, pig, rabbit etc) has been determined. Interestingly, it has been observed that the sequence in the A- chain remains invariant except for three amino acid residues numbered 8, 9 and 10, which are Thr, Ser and Ile respectively for the human insulin. On the other hand the B-chain of the insulin from the elephant is identical to that of the human while in case of pig, sheep, whale horse and cattle etc. the Thr residue at position number 30 is replaced by Ala. The structure of insulin has been confirmed by its synthesis. Since insulin is a very important molecule for the treatment of diabetic patients, a great deal of research is devoted to get human insulin from other sources. A semi-synthetic version of human insulin has been produced by a method called transpeptisation.

2.6.3 Bradykinin

Bradykinin is a nonapeptide (peptied containing nine residues) which is released by the blood plasma in response to a wasp sting. Physiologically bradykinin is a pain causing agent having the following amino acid sequence.

Arg–Pro–Pro–Gly–Phe–Ser–Pro–Phe–Arg

Bradykinin is also known as Kalladin I which has a closely related peptide Kalladin II (a decapeptide) with the following sequence.

Lys–Arg–Pro–Pro–Gly–Phe–Ser–Pro–Phe–Arg

Both the Kalladins have excellent muscle relaxing properties.

2.6.4 Gramicidin S

Gramicidin S is a peptide antibiotic isolated in 1944. This cyclic decapeptide contains two identical pentapeptides viz., D-Phe-L- Pro-L-Val-L-Orn-L-Leu. The two pentapeptides are condensed head to tail with each other. Ornithine is an amino acid similar to lysine with one methylene group lesser. The structure of the peptide is as follows:

Val-Orn-Leu-Phe-Pro
| |
Pro-Phe-Leu-Orn-Val

Gramicidin S exhibits a broad spectrum antibiotic activity, which is attributed to its ability to bind cell membranes. However, its clinical applications are quite limited due to its low bioactivity and cytotoxicity. The three dimensional structure of the peptide, co-crystallised with urea is shown to contain a double stranded antiparallel β-sheet structure formed by two similar tripeptides (Val-Orn-Leu) stabilised by four hydrogen bonds as shown in Fig 2.9.

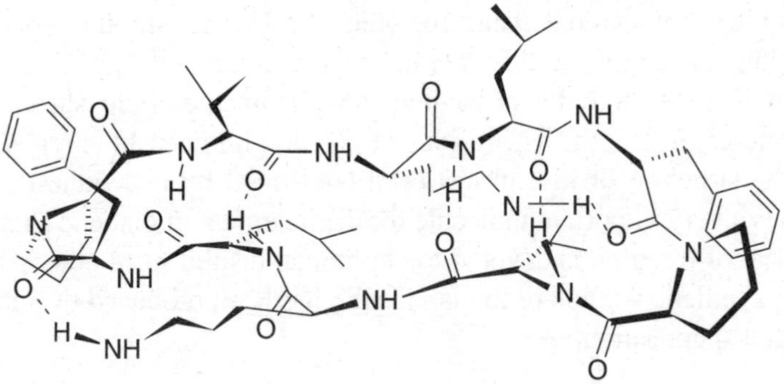

Fig. 2.9 *The three-dimensional structure of gramicidin S.*

The structure and functions of some biologically active peptides are compiled in Table 2.1

Table 2.1 Structures and functions of some biologically active peptides

Name of the peptide	Primary structure (Amino acid sequence)	Biological functions
Angiotensin II (horse)	1 8 H-Asp-Arg-Val-Tyr-Ile-His-Pro-Phe-OH	Hypertensive peptide; controls blood pressure by constriction of arteries. Also involved in release of aldosterone from adrenal gland
Glucagon (bovine)	1 10 H-His-Ser-Gln-Gly-Thr-Phe-Thr-Ser-Asp-Tyr- 11 20 Ser-Lys-Tyr-Leu-Asp-Ser-Arg-Arg-Ala-Gln- 21 29 Asp-Phe-Val-Gln-Trp-Leu-Met-Asn-Thr-OH	Pancreatic hormone involved in regulation of glucose metabolism
Plasma bradykinin	1 9 H-Arg-Pro-Pro-Gly-Phe-Ser-Pro-Phe-Arg-OH	A vasodilator
Enkephalin (Met-enkaphalin)	1 5 H-Tyr-Gly-Gly-Phe-Met-OH	A member of family of opiate-like peptides found in brain that inhibit sense of pain by binding to receptors in brain

Substance P	1
	H-Arg-Pro-Lys-Pro-Gln-Phe-Phe-
	                         10
	                  -Gly-Leu-Met (NH$_2$)
		A neurotransmitter

Thyrotropin releasing factor (TRF)	1 3 pyroGlu-His-Pro(NH$_2$)*	Secreted by hypothalamus and causes pituitary gland to release thyrotropin (or thyroid-stimulating hormone, TSH).

Vasopressin (antidiuretic hormone)	1 9
	H-Cys-Tyr-Phe-Gln-Asn-Cys-Pro-Arg-Gly(NH$_2$)
	   | |
	   S————————S
		Secreted by pituitary gland and causes kidney to reabsorb water from urine

EXCERCISES

1. What is a peptide bond? Discuss the structural features of the peptide bond. What significance does it have in the structure of proteins?

2. Peptides are very versatile biological molecules. Comment.

3. Met-Enkephalin - an opiod peptide has the following sequence

    Tyr-Gly-Gly-Phe-Met

    (a) Express the peptide in terms of one letter codes of the amino acids.
    (b) Write the structure of the peptide (you may use Table 1.1) and mark its N-terminal and the C-terminal.
    (c) Predict the net charge on the peptide at a pH of 6.0.

4. What do you understand by protection of amino acids? What essential criteria need to be fulfilled by the protecting groups in peptide synthesis?

5. List all the possible tetrapeptides having a composition of Ala, $(Gly)_2$, Val.

6. Write the chemical equations with appropriate reaction conditions for the preparation of following protected amino acids.

    (a) Boc-Ala
    (b) Fmoc-Val
    (c) Cbz-Gly
    (d) Leu-OMe
    (e) α, ϵ- di Boc-Lysine

7. Outline a suitable synthetic procedure for the solution phase synthesis of the tripeptide, Ala-Gly-Val.

8. How would you prepare poly-L- alanine from alanine using Leuch's anhydride method? Give detailed reaction steps involved.

9. What is solid phase peptide synthesis? Discuss its advantages and limitations.

10. Outline a strategy for synthesising a tripeptide Ala-Lys-Ala using Merrifield procedure.

11. Discuss the biological role of the peptide glutathione.

12. List salient features of the structure of insulin.

Chapter 3

Proteins

3.1 INTRODUCTION

Proteins are a class of biopolymers which probably affect every aspect of living organisms. The term protein originates from the Greek word 'προτειοσ' (*proteios*, meaning 'of first importance') and was coined by Dutch chemist G.T.Mulder in 1839 on the suggestion of Swedish chemist Jons J Berzelius. Some authors ascribe the origin of the word 'protein' to Latin word '*primarius*' or from Greek god '*Proleus*'. Proteins represent an enormous group of complex nitrogenous molecules widely present in plants and animals. These perform a wide range of functions like providing structure to the body, transporting oxygen and other substances within an organism, regulating body chemistry etc. in the living systems. Skin, bone, horn, tendon, muscle, feather, tooth and nail are all proteinaceous in nature. The cellular membranes contain proteins which control their permeability and allow the selective solute molecules to pass through against thermodynamic gradients. The biochemical regulation is performed by hormones, majority of which are proteins or peptides. Our body depends on tens of thousands of proteins to perform its normal functions and these proteins differ from one another in their composition and structure.

Selectively toxic peptides and proteins are deployed by many living species for their defense, for example, the venoms of snakes, bees and wasps and the lethal bacterial toxin responsible for botulism etc. Our immune system employs proteins to selectively recognise and reject foreign molecules, thus exercising precise discrimination between self and non-self. Proteins are also required in the genetic expression and control. Though the genetic information is encoded in nucleic acid structures, a number of proteins like, histones (the basic proteins that keep DNA in highly organised compact structure) are required for translating this information.

Proteins owe their structural and functional diversity to different combinations of about twenty amino acids that join through peptide bonds to make these large polymeric molecules. Different proteins differ only in their composition and the sequence in which the constituent amino acids are linked together. The functions and properties of proteins are intimately related to their structure which has different levels of organisation. In this chapter we are going to focus our attention on an understanding of the protein structure and its determination.

3.2   CLASSIFICATION OF PROTEINS

Proteins show a wide range of structural and functional diversity. It is quite difficult to classify this diverse group of biological macromolecules on the basis of one single property or characteristic. Thus, there is no general or universal system of classification of proteins. Some of the common basis of classification of proteins are given below and are discussed subsequently.

- Shape and structure
- Products of hydrolysis
- Biological function

3.2.1 Classification on the Basis of Shape and Structure

Proteins adopt a wide range of shapes depending on their composition and the nature of folding of polypeptide chains. On the basis of their overall shape the proteins are further categorised into two broad classes. These are

- Globular proteins
- Fibrous proteins

Globular Proteins

These are also called **spheroproteins** or **corpuscular proteins** and contain compactly folded coils of polypeptide chains which give them a shape of spheroids (a sphere) or ellipsoids (an ellipse). These have low **axial ratio** (<10, usually 3-4) i.e., the ratio of length to the width. These proteins contain a number of structural units, called domains which are of single strand polypeptide chains containing alternating α-helices and β-strands (structures adopted by proteins; discussed later). These are usually soluble in water or aqueous solutions of acids, bases or alcohols. Due to the compact shape these proteins diffuse quite easily. Albumins, globulins, lysozyme, histones and protamines - the proteins of animal origin and prolamines and glutelins found in the food grains belong to this class.

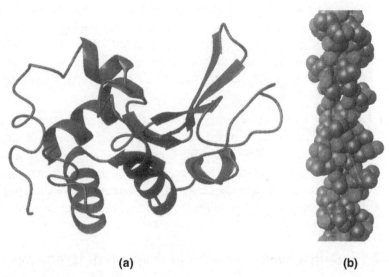

(a)   **(b)**

Fig. 3.1 *(a) Lysozyme–a globular protein and (b) Collagen–a fibrous protein. The globular proteins have low axial ratios as compared to the fibrous proteins.*

Fibrous Proteins

The proteins belonging to this class have an elongated shape and look somewhat like fibres or threads and have axial ratio of more than 10. These are insoluble in water and aqueous solutions of acids and bases. Fibrous proteins have high mechanical strength and form skeletal and connective tissues in animals. The peptide chains in these proteins can be arranged in many ways, and depending on these, fibrous proteins can be further divided into following special types.

(i) **α- Keratins:** α-Keratins are found in protein fibres of hair, skin and wool and are constituted of a number of α-helices that are twisted together like the strands of a rope. These are rich in cystine and poor in proline and hydroxyproline. Fibrin, the protein involved in the coagulation of blood also has a α-keratin structure.

(ii) **β-Keratins:** β-Keratin type of fibrous proteins are found in hard tissues such as nails, horns, spider's web, fish scales, porcupine quills and bird feathers etc. These are rich in small uncharged amino acids like glycine and alanine and poor in cysteine, proline and hydroxyproline and acquire a β-pleated sheet structure. Silk fibroin obtained from the cocoons of silk moth has β-keratin structure.

(iii) **Collagens:** Collagens are the main components of bone, teeth, connective tissues of tendons and ligaments etc. These are rich in amino acids like glycine, alanine, proline and hydroxyproline and poor in cysteine, cystine, methionine, tyrosine and tryptophan. In collagen, the polypeptide chain acquires an elongated left handed α-helical structure and three such helices spiral around each other like the strands of a rope. On heating with water these proteins swell up to give colloidal gelatin.

(iv) **Elastins:** These are closely related to collagens and constitute ligaments, the walls of blood vessels etc. Elastins are rich in amino acids like, alanine, leucine, proline and valine and are deficient in cysteine, cystine, methionine and histidine. Structurally, elastin is similar to collagen, the difference being in the nature of cross linking of different polypeptide chains. Further, it does not swell up in water like collagen.

3.2.2 Classification on the Basis of Products of Hydrolysis

On the basis of the products obtained on hydrolysis (or the composition) proteins can be classified into three categories as:

- Simple proteins
- Conjugated proteins
- Derived proteins

Simple Proteins

These are composed of amino acids only and do not contain any non-protein part attached to them. These are further classified on the basis of their solubility, precipitability and thermocoagulability which in turn depend on the size and shape of the proteins. Most of the simple proteins are of globular type, scleroproteins being exception as these are fibrous in nature. Major classes in this type include the following.

(i) **Albumins:** These are soluble in water and aqueous solutions of acids, bases and salts. Albumins generally have low isoelectric point; pI and act as acidic proteins under physiological conditions. These can be reprecipitated by full saturation with ammonium sulphate and are easily thermocoagulable. These are generally deficient in glycine. Serum albumin in blood plasma, ovalbumin in egg white and lactoalbumin in milk are common animal albumins while leucosin in cereals, legumelin in legumes and ricin in castor beans are common plant albumins.

(ii) **Globulins:** These are ellipsoidal in shape, having lower solubilities as compared to albumins and are insoluble in water but dissolve in aqueous solutions of acids, bases and dilute salt solutions (5% sodium chloride).

These can be precipitated by half-saturation with ammonium sulphate and are also easily thermocoagulable. These usually contain the amino acid, glycine. Serum globulin in blood plasma, myosin in muscles and lactoglobulin in milk and thyroglobulin from the thyroid gland are common animal globulins while edestin in hemp seed is an example of plant globulin.

(iii) **Histones:** These are a small group of closely related basic nucleoproteins (proteins found in the nucleus of the cell) containing a good fraction of lysine, histidine or arginine and lack tryptophan and are deficient in cysteine and methionine residues. These can be isolated from the sperm cells, red blood corpuscles (RBCs) and white blood corpuscles (WBCs). Histones dissolve in water and dilute aqueous solutions of acids and bases but are insoluble in dilute ammonia. These are not easily thermocoagulable.

(iv) **Protamins:** Protamins, like histones, are also basic proteins that are soluble in water and dilute aqueous solutions of acids, bases and ammonia and are not easily thermocoagulable. These occur almost entirely in animals and are the main components of sperm cells of certain fishes. Salmine and clupeine are the protamins found in the sperm cells of salmon and herring fish respectively. The protamins are quite small in size and have relatively low molecular mass of the order of ~5000 Daltons. These proteins are quite rich in basic amino acid, arginine and behave more like a polypeptide.

(v) **Prolamins:** This is a very small class of plant proteins found in grain stuff. These proteins are insoluble in either water or alcohol and are not thermocoagulable but can be extracted with 60-80% alcohol or dilute aqueous solutions of acids and bases. These are rich in glutamic acid and proline but poor in histidine, lysine and arginine etc. Gliadin from wheat, hordenine from barley and zein from maize etc. are common examples of prolamins.

(vi) **Glutelins:** These, like prolamins are plant proteins found in grains. These are insoluble in either water or alcohol but can be extracted with dilute aqueous solutions of acids and bases. Due to their reasonable size these can be thermocoagulated. It is in contrast to prolamins which are not thermocoagulable. Glutelins in a way supplement prolamins as these contain lysine and tryptophan. Glutenin in wheat flour, glutelin in corn and oryzenin in rice are common examples of glutelins.

(vii) **Scleroproteins:** Scleroproteins are fibrous proteins of animal origin. These are insoluble in water and dilute aqueous solutions of acids, bases and salts and also in 60-80% alcohol. However, these dissolve in concentrated acids and bases and include α- and β- keratins, collagens and elastins discussed above.

Conjugated or Complex Proteins

When the proteins form complexes with some nonprotein components these are called conjugated proteins. The protein part of conjugated proteins is called **apoprotein** and the nonprotein part is referred to as the **prosthetic group**. The entire molecule is called a **haloprotein**. The conjugated proteins are generally globular in nature and can be further subdivided on the basis of the nature of prosthetic group present. Major classes in this type include the following.

(i) **Phosphoproteins:** These proteins contain phosphoric acid as prosthetic group and do not include phosphate containing substances like nucleic acids etc. The phosphoric acid group is connected through ester linkage to a hydroxy group on the polypeptide chain. Casein, the milk protein and ovovitellin obtained from egg yolk are common phosphoproteins. The phosphoric acid group of these proteins can be removed enzymatically or with the help of aqueous sodium hydroxide.

(ii) **Metalloproteins:** In metalloproteins, the protein molecule is linked with metal ions which may be strongly bound or weakly associated. Siderophilin, an important plasma protein, constituting about a third of plasma proteins, has a very strong affinity for iron and in fact is involved in iron transport. Ceruloplasmin, another plasma protein has affinity for copper ions. In addition, a number of enzymes also need metal ions for their activity. For example, *carbonic anhydrase* requires Zn^{2+} while *cytochrome oxidase* needs Cu^{2+} and Fe^{3+} ions to perform their functions.

(iii) **Chromoproteins:** Chromoproteins are proteins containing coloured pigments like flavins, carotenoids, porphyrins etc. as their prosthetic groups. This class includes proteins involved in respiratory functions e.g., haemoglobin, myoglobin and cytochrome etc. A number of enzymes like, catalase, peroxidase, cytochrome oxidase belong to this group. Chloroplastin, the plant protein containing chlorophyll is also a chromoprotein.

(iv) **Glycoproteins and mucopolysaccharides:** This group refers to the proteins containing a carbohydrate component as the prosthetic group. The prosthetic group is an integral part of the structure and may constitute from less than 4% to more than 80% of the glycoproteins. The oligosaccharide chains are joined with the polypeptide through a

O-glycosidic linkage with the hydroxyl group of serine or threonine residues or by N-glycosidic linkage with the amide side chain of asparagine. Glycoproteins include immunoglobulins, many enzymes and some hormones like, thyrotropin or thyroid stimulating hormone (TSH), follicle stimulating hormone (FSH) and lutenising hormone(LH) etc. These hormones are secreted by anterior pituitary glands. **Mucopolysachharides**, on the other hand, contain a small protein component in a carbohydrate structure. These are found in the supportive and connective tissues in animals. Hyaluronic acid present in the vitreous humour of the eye is one of the simplest mucopolysachharides and is made up of N-acetylglucosamine and D-glucoronic acid. Heparin, the natural anticoagulant and chondroitin sulphate isolated from cartilages and tendons are some of the other examples of mucopolysaccharides.

(v) **Nucleoproteins:** Nucleoproteins are complexes containing nucleic acids and basic proteins like protamines and histones. The protein and the nucleic acids are bound primarily through electrostatic interactions. These complexes adopt very compact structures and are important constituents of chromosomes, ribosomes, and some viruses.

(vi) **Lipoproteins:** This group includes proteins complexed with lipids and have a variable composition. The nonpolar side chains of the apoprotein and the lipids are bound through hydrophobic interactions. These are important constituents of blood plasma, cellular membranes, milk and egg yolk etc. Lipoproteins are further classified on the basis of their densities which in turn depend on the protein content. The lipoproteins of lowest density are called as chylomicrons and contain up to 2% protein. The fraction containing about 9% of the proteins is referred to as **very low density lipoproteins (VLDL)**; the one having about 21% protein are called **low density lipoproteins (LDL)** while the lipoproteins containing about 33% proteins are known as **high density lipoproteins (HDL)**. These lipoprotein complexes are closely related to heart diseases. Increased concentration of LDL in the blood increases the risk of **arteriosclerosis** –the thickening of arterial walls.

Derived Proteins

These are the degradation products derived on subjecting the native proteins to different physical or chemical agents like, heat, acid, alkali or enzyme. Artificially produced polypeptides are also included in this class. The derived proteins are further classified into following types.

(i) **Primary or denatured derived proteins:** These are obtained by the action of water, acids, alkalies, heat, radiations etc. on the native protein which denature the protein without hydrolysing it. It implies that the perturbing conditions do not hydrolyse the peptide backbone, only the weaker forces like hydrogen bonding, hydrophobic interactions, salt linkages etc. responsible for maintaining the native structure of protein are disrupted. In these, proteins molecular mass does not change, only the properties like solubility, precipitation etc. are altered e.g., ovalbumin, a soluble globular protein changes to insoluble fibrous protein on heating or treating with urea.

(ii) **Secondary derived proteins:** These are obtained by the processes that cause progressive hydrolysis of the native protein. The products obtained in the initial stages of the hydrolysis, with dilute acids, alkalies and enzymes, are called **proteoses**; these are soluble in water and are heat coagulable. The primary proteoses also called **metaproteins** are insoluble in water but dissolve in acids and alkalis. These can be salted out by half saturation with ammonium sulphate. However, the **secondary proteoses**–the second stage hydrolysis products need full saturation. The next stage of hydrolysis generates **peptones** which are soluble in water and are not coagulated by heat and also cannot be salted out by ammonium sulphate. Further hydrolysis gives polypeptides and simple peptides, which also behave like peptones. The ultimate product of the hydrolysis is a mixture of amino acids. These are soluble in water and sparingly soluble in organic solvents. The products obtained at different stages of protein hydrolysis and their properties can be summarised as follows:

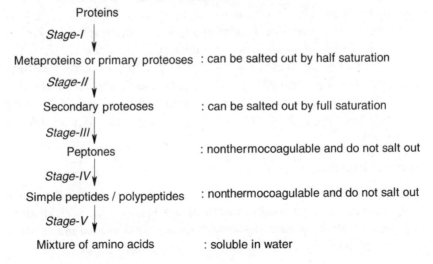

Proteins

Stage-I ↓

Metaproteins or primary proteoses : can be salted out by half saturation

Stage-II ↓

Secondary proteoses   : can be salted out by full saturation

Stage-III ↓

Peptones   : nonthermocoagulable and do not salt out

Stage-IV ↓

Simple peptides / polypeptides : nonthermocoagulable and do not salt out

Stage-V ↓

Mixture of amino acids   : soluble in water

3.2.3 Classification on The Basis of Biological Functions

Since proteins perform a variety of functions in the body, the biological functions performed provide another basis for their classification.The different classes of proteins on the basis of their biological function are as follows.

(i) **Enzymes:** These are also called as biochemical catalysts and are proteins involved in catalysing biological reactions e.g., *dehydrogenases*, catalysing dehydrogenation reactions and, *kinases*, involved in transfer of phosphate group etc. Chapter 4 gives a detailed account of enzymes in terms of their structure and function etc.

(ii) **Storage proteins:** These proteins are entrusted with the job of storing important species in the living cell. Myoglobin (responsible for oxygen storage in skeletal muscles) and ferritin (involved in the storage of iron in the liver) belong to this category.

(iii) **Regulatory proteins:** As the name suggests, these proteins control many aspects of cell functions including metabolism and reproduction. Several hormones such as insulin and glucagons that regulate body functions are also proteins.

(iv) **Structural proteins:** These provide mechanical support to large animals and provide them with their outer covering, e.g. proteoglycan. These proteins are also major structural components of cellular membranes and cell organelles. Our hair and finger nails are largely composed of keratin, a type of fibrous protein. Collagen - another fibrous protein provides mechanical strength to our teeth, bones and tendons and is responsible for the stability of our body.

(v) **Protective (or defense) proteins:** These proteins defend the body against infection and are related to the immune system. These are collectively called as **immunoglobulins** or antibodies which bind specifically to an antigen (foreign substances / organisms causing infection) and cause its destruction. Venoms and toxins also belong to this class.

(vi) **Transport proteins:** These proteins are engaged in carrying materials from one place to another in the body. These are generally involved in transport of food molecules, energy sources etc. across the cell membrane or in the body fluids. For example, haemoglobin is associated with the transport of oxygen while the protein called transferrin is used for transporting ferric ions in the plasma.

(vii) **Contractile and mobile proteins:** These are engaged in all forms of movement for example, actin and myosin are responsible for heart and muscle movement, tubulin for sperm movement etc.

3.3 PROPERTIES OF PROTEINS

Proteins exhibit properties generally shown by the constituent amino acids. However, because of their polymeric nature they show some other typical physical and chemical properties also as outlined below.

3.3.1 Molecular Weight

Proteins vary tremendously in size, from say of only 50 amino acid residues to the ones consisting of as many as 25000 amino acid residues or even more. Further, these may exist as individual chains or as aggregates of a number of copies of same polypeptide or as one or many copies of different polypeptides. These may have molecular weights (or molecular mass) ranging from about 6000 to 3000000 units. The molecular weight of proteins expressed in the units of Daltons (Da)- a unit of mass defined to be equal to 1 a.m.u which is almost equal to the mass of a hydrogen atom. In protein biochemistry the term, 'molecular weight' though incorrect is widely used in place of molecular mass. The largest known protein is human titin, associated with arrangement of repeating structure of muscle fibres. Titin contains 26926 residues and has a molecular mass of 2990 000 Daltons. However, most of the proteins contain between 100-1000 residues. The number of amino acid residues and the molecular masses of some of the proteins are given in Table 3.1.

Table 3.1 Sizes of some common proteins

Protein	Number of amino acid residues	Molecular mass (Da)
Cytochrome C	104	13 000
Ribonuclease H	155	17 600
Triose phosphate isomerase	510	56 400
Haemoglobin	574	64 500
Pyruvate decarboxylase	2252	250 000
Titin	26926	2 990 000

The molecular mass of proteins is determined by molecular sieve chromatography (gel filtration), the sedimentation analysis, osmotic pressure, freezing point depression, light scattering, x-ray diffraction, turbidity or by newly developed technique of electron spray ionisation mass spectrometry etc. The proteins also show colloidal properties due to their high molecular masses. Such proteins are found to be highly viscous in nature.

Sometimes, the native protein may be associated with nonprotein component which may be large chains of carbohydrate or as mentioned earlier the protein itself may be a combination of a number of polypeptides. These combinations or aggregates alter the molar mass significantly. It is therefore advisable to subject the protein to a strong denaturant like 6 M guanidinium hydrochloride so as to dissociate the aggregate if present. The individual polypeptides are then separated and their molecular weight determined. The experimentally determined molecular weight of a polypeptide on dividing by 120 (the weighed average molecular weight of amino acid residues) gives an estimate of the number of amino acid residues in the protein.

3.3.2 Amphoteric Nature

Most proteins contain a number of ionisable groups in the side chains of the amino acid residues like lysine, glutamic acid and arginine etc., in addition to the terminal amino and carboxyl groups of the amino acids constituting them. These ionisable groups can act as proton donors or acceptors depending on the pH of the medium. Thereby, these groups impart an amphoteric character to the protein. Different groups ionise to variable extents depending on the pH, nature of neighbouring side chains and the location of the group in the overall three dimensional structure of the protein i.e., whether the groups are on the surface of the protein or are buried inside it. Thus the proteins carry a number of positive and negative charges which changes on changing the pH of the medium. At a certain pH (referred to as **pI, isoelectric point**), the positive and negative charges balance each other and the protein becomes neutral i.e., the net charge on the protein is zero. In the presence of applied electric field such a protein would not move towards any of the electrodes. This fact is exploited in the purification or separation of different proteins in what is called as **gel electrophoresis**.

Some proteins e.g., histones are rich in basic amino acids like lysine, arginine and histidine etc., containing cationic side chains. The pI of such proteins lie in the alkaline range and at physiological pH of 7.4, these proteins behave as bases and are called **basic proteins.** On the other hand proteins containing a major portion of anionic side chains like that of aspartic and glutamic acids etc. would have their pI in the acidic range. At physiological pH, these proteins behave as acids and are called as **acidic proteins.**

The zwitterionic form of protein is responsible for its high dipole moment and high dielectric constant. It is because of the zwitterionic structure that in a crystal lattice also protein molecules are held together tightly. As a result, proteins have high melting points.

3.3.3 Solubility

Solubility of protein molecules, in general depends upon their structure and nature besides pH, temperature, polarity of solvent and concentration of dissolved salts etc. The globular proteins are more soluble because of their compact folded structures and presence of polar side chains on the surface. These polar side chains being hydrophilic interact favourably with water and contribute to the solubility. At the isoelectric pH proteins have very low or zero net charge which eliminates the mutual repulsion amongst different molecules of protein. As a consequence protein molecules come together to aggregate and the solubility is reduced to a minimum. On adding acids or alkalies, the pH changes to either side of the pI where the molecules are charged and the solubility increases. Thus, the solubility of a protein is minimum at its isoelectric point and increases as the pH is decreased or increased. The isoelectric points of some common proteins are given in Table 3.2.

Table 3.2 Isoelectric points of some common proteins

Protein	Isoelectric point
Lysozyme	11.1
Lactoglobin	5.2
Cytochrome C	10.7
Urease	5.0
Thymohistone	10.6
Gelatin	4.8
Chymotrypsin	9.5
Serum albumin	4.7
Ribonuclease	7.8
Casein	4.6
Haemoglobin	7.2
Ovalbumin	4.6
Fibrinogen	5.8
Globulin	2.0
α_1- Lipoprotein	5.5
Pepsin	ca. 1.0
Insulin	5.4

Further, the proteins are more soluble in solvents with high dielectric constants like water while the solubility is low in solvents like alcohol, chloroform and benzene which have a low dielectric constant. In water, the polar groups of proteins bind to the solvent molecules by H-bonds as a consequence of which the protein molecule is surrounded by water molecules. These water molecules form a kind of cage around the protein molecules and help them stay in solution. Addition of solvents of low dielectric constant like alcohol to an aqueous solution of the protein decreases the dielectric constant of the medium. This decreases the cage effect and thereby leads to aggregation or precipitation of the protein.

3.3.4 Precipitation

As discussed above, every protein has a net positive or negative charge on it. The interaction of these charges with ions and water are responsible for the solubility of proteins. Repulsion between these charges keeps the protein molecules away from each other. If these charges are somehow neutralised, the protein molecules would come close, aggregate and result into precipitation. There are different ways in which precipitation of proteins can be achieved.

As discussed earlier, at their pI, the protein molecules are in their zwitterionic form and carry minimum charges and can easily get aggregated and precipitate out without denaturation. Thus the adjustment of the pH of a protein solution to its pI can lead to the precipitation of the protein. This is called **the isoelectric precipitation**. In isoelectric precipitation, pH of the solution is adjusted to the pI of the protein by adding suitable quantities of acid or a base.

The ions of heavy metals like, Hg^{2+}, Ag^+ and Pb^{2+} can form linkages with the acidic groups in the side chains of the amino acids like, glutamic acid and aspartic acid and the sulphur atoms in disulphide linkages. This causes neutralisation of the negative charge on the protein leading to its coagulation. It is for this reason that the salts of heavy metal ions act as poisons, if taken internally. Interestingly, the antidote for such a poisoning is also based on the same action. Raw egg white or milk is given as an antidote for heavy metal poisoning.

Polar organic solvents like ethanol and acetone are capable of forming hydrogen bonds with proteins. In the process of making such bonds they break protein's normal hydrogen bonding pattern with the solvent water. As the water structure around the protein collapses, the protein coagulates. A 70% ethanol solution acts as a disinfectant as it causes the bacterial proteins to denature leading to the death of the bacterial cell.

Addition of concentrated solutions of neutral mineral salts as $MgSO_4$, Na_2SO_4 or $(NH_4)_2SO_4$, etc. to the aqueous solution of the protein disrupts the water cage around the protein molecules and facilitates its aggregation. The solubility of proteins in aqueous solution containing low salt in fact increases on adding salt. This is due to the shielding of the charges on the protein molecules and thereby keeping them apart. This is referred to as **salting in**. Further addition of salt in excess reduces the charge on protein which again helps in the aggregation process. This process of separation of protein from its solution on adding high concentrations of mineral salts is called **salting out** or **coagulation** of protein. Ammonium sulphate is the reagent of choice for the purpose because its solubility is very high and a solution of over 4 M can be prepared at room temperature.

3.3.5 Denaturation

Proteins may undergo disruption of their overall structure when subjected to heat, X-rays, ultraviolet rays, high pressure, concentrated mineral acids, acetone, alcohols, urea, heavy metal ions, sodium dodecyl suphate, detergents, violent agitation and dilution etc. These perturbants cause the structural change by rupturing the noncovalent interactions between the side chains of different amino acid residues responsible for the stability of secondary and tertiary structures of proteins. This change in structure is called the **denaturation** of protein molecules. The process of denaturation, however, does not change the primary structure of proteins i.e. the amino acid sequence remains the same.

Denaturation is generally an irreversible process, however in some cases like, denaturation of pancreatic ribonuclease by adding high concentration of urea, it can be made a reversible process. The native protein can be regenerated by gradually removing the denaturing agent and the process is called **renaturation**. For any denaturation to be reversible the native protein must have a lower free energy than the denatured form.

3.3.6 Colour Reactions

Certain reagents react with some specific amino acids (tyrosine, tryptophan etc.) or groups (e.g. peptide bond) in the protein molecule to give coloured products. These reactions act as the basis for qualitative detection of proteins and are discussed below.

(i) **Biuret reaction :** The biuret reaction is in fact a test for the presence of a peptide bond (–CONH–). In this test biuret reagent (copper (II) sulphate solution in alkaline tartarate) reacts with a molecule containing at least two peptide bonds to give a violet/purple coloured complex that has maximum absorbance at 540 nm. The exact nature of the complex is uncertain, it is likely to be a coordination complex between peptide bonds and cupric ion. Since proteins contain a large number of peptide bonds, these give a positive biuret test. Peptides, other than dipeptides, also give a positive response to this test. However, since the complex formation invokes the involvement of peptide NH group, the proteins rich in proline residue will give a poor response because it cannot participate in this reaction.

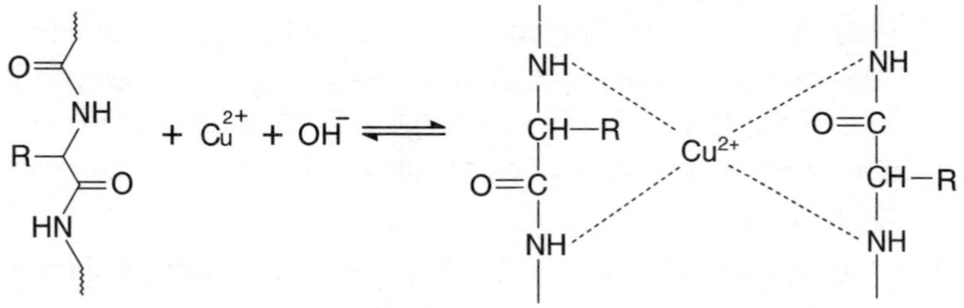

This reaction can also be used for quantitative determination of the proteins. For this, the absorbance of the protein solution (of unknown concentration) in presence of the reagent is compared with calibration curve obtained with a standard protein. BSA (bovine serum albumin) is commonly employed for calibration purposes.

(ii) **Xanthoproteic reaction:** This reaction is due to the presence of aromatic amino acids like, tyrosine or tryptophan in the protein molecule. In this test addition of concentrated nitric acid to the protein gives a yellow colouration, which turns orange on adding sodium or ammonium hydroxide. This reaction is responsible for the yellow colouration of skin when some nitric acid falls on it.

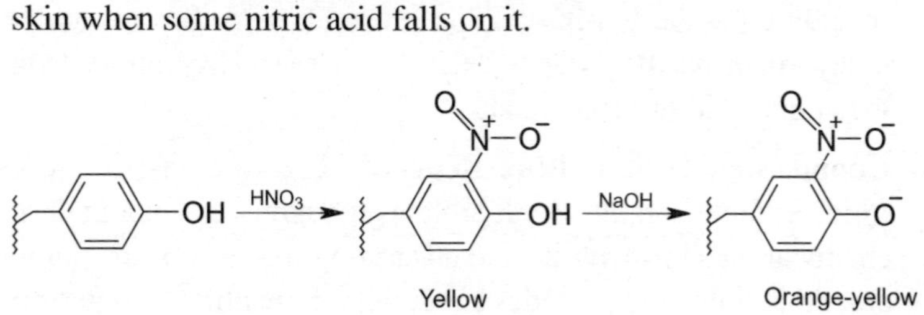

Yellow   Orange-yellow

(iii) **Millon's reaction:** This reaction is due to the presence of amino acid, tyrosine, in the protein molecule. Addition of Millon's reagent (mercuric sulphate in sulphuric acid) to protein solution and heating followed by addition of sodium nitrite gives a red colouration. Since this reaction is characteristic of phenols even the nonprotein phenolic molecules may also give a positive test with Millon's reagent.

(iv) Ninhydrin reaction: Ninhydrin is probably the most extensively used reagent for the detection of proteins and amino acids. Ninhydrin reacts with free amino groups present in the protein or the ones generated by hydrolysis of the protein. As discussed earlier (Sec.1.5.3) **ninhydrin** causes oxidative deamination of the amino acid and gives an intense violet coloured product. This coloured product has a λ_{max} of 570 nm. Proline and hydroxyproline however, give a different product (yellow) of an uncertain structure having a λ_{max} of 440 nm. This reaction forms the basis of detection of amino acids in their chromatographic determination wherein ninhydrin acts as a spray or detection agent. The ninhydrin reaction however has a disadvantage that in the process of detection the amino acid or protein is destroyed and cannot be recovered.

(v) Fluorescamine reaction: Fluorescamine is a very sensitive reagent for the detection of amino acids. Like ninhydrin, this reagent also reacts with the amino group of amino acids and proteins but it gives a product that is fluorescent in nature.

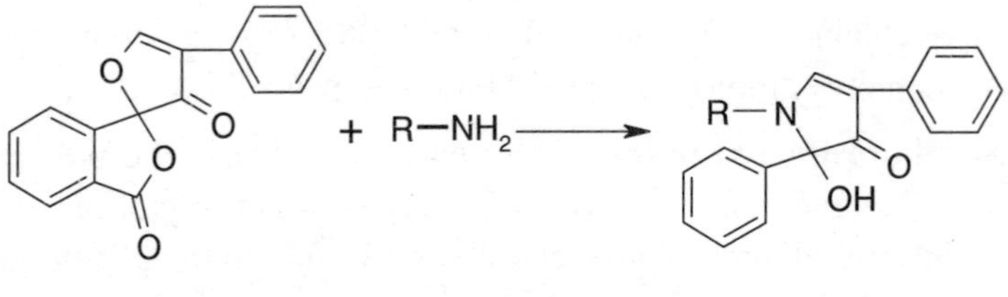

Fluorescamine

In addition to the sensitivity, this reagent has another advantage over ninhydrin in that the product obtained can be acid hydrolysed to recover the amino acid or protein back.

(vi) Coomassie Brilliant Blue Reagent: The Coomassie dye G 250 forms a stable complex with proteins though it does not undergo any chemical reaction with it. The absorption maxima of the complex is around 595 nm and provides a method for quantitative determination of proteins. It is a general method in which the complex formation seemingly involves nonpolar interactions and does not rely on a few side chains or groups in the protein.

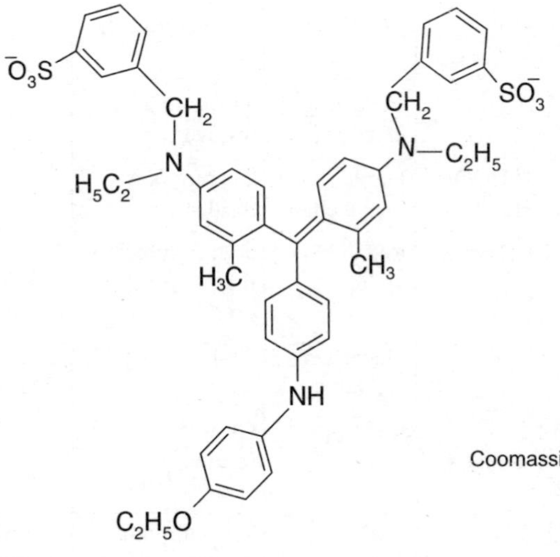

Coomassie dye G 250

(vii) **Ortho-phthalaldehyde reaction:** Ortho-phthalaldehyde is another reagent that gives a fluorescent product with amino group of amino acid or protein. The reagent reacts with the amino group in the presence of mercaptoethanol.

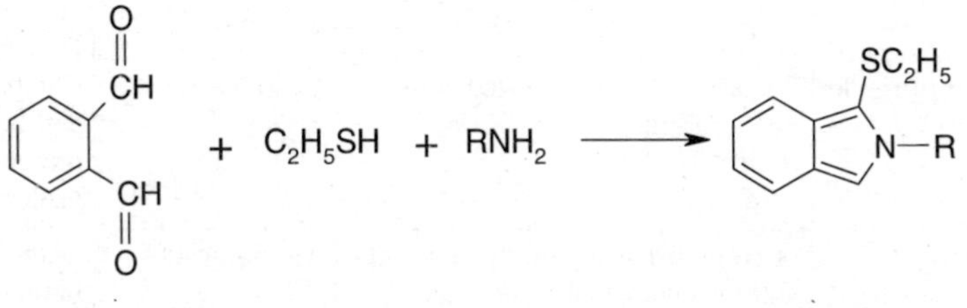

o-Phthalaldehyde

The reaction is sensitive to the extent that it can detect pico mol (10^{-12} mol) quantities of amino acids or proteins.

In addition to the above discussed coloured reactions, proteins show a number of other colour reactions. The most common colour reactions of proteins are compiled in Table 3.3.

Chemistry of Natural Products

Table 3.3 Summary of some coloured reactions of proteins

Name of the reaction/test	Reagent	Colour	Reaction site	Remark
Acree-Rosenheim	Dil HCHO and conc H_2SO_4	Violet ring at junction	Tryptophan residue	—
Adamkiewez	Glacial CH_3COOH and conc H_2SO_4	Purple ring at the junction	Tryptophan residue	—
Biuret	Dil $CuSO_4$+NaOH	Red or blue-violet	Two or more than two peptide bonds	Not given by amino acids or dipeptides
Fluorescamine	Fluorescamine	Fluorescent derivative	Amino groups	Detected by fluorescence
Folin's reaction	Sodium 1,2-naphthaquinone-4-sulphonate	Red	Cysteine residue	—
Hopkins- Cole	Gloxylic acid and conc H_2SO_4	Violet or purple ring	Trypotophan residue	Interference from compounds containing indole ring
Millon's	Mercuric sulphate in sulphuric acid and sodium nitrite	Red precipitate	Phenolic groups, tyrosine	Also given by phenols
Ninhydrin	Ninhydrin	Blue violet absorbing at 540 nm	Amino groups	Proline and hydroxyproline give yellow colour
Nitroprusside	Sodium nitroprusside in dil NH_4OH	Red colour	Cysteine residue	Can be used to detect the presence of 'S' in proteins
Ortho-pthalaldehyde	ortho-phthalaldehyde and mercaptoethanol	Fluorescent derivative	Amino groups	Detected by fluorescence
Pauly's reagent	Diazotised sulphanilic acid in alkaline solution	Yellow to red product	Tyrosine (phenol) or histidine (imidazole)	—
Sakaguchi	Sodium hypochlorite and α-naphthol	Red	Arginine residue	Interference from compounds containing guanidine
Xanthoproteic	Boiling with HNO_3	Yellow or orange	Phenolic or indolic amino acids, e.g., phenylalanine, tyrosine	Responsible for yellowing of skin when HNO_3 falls on it

3.4 STRUCTURAL ORGANISATION OF PROTEINS

Proteins show a great deal of functional diversity. This can be attributed to the diversity in the chemical nature of the amino acids constituting them and the variety of the structures adopted by the protein molecules. Since the properties of the proteins are intimately related to their structure, it is imperative to have an understanding of the structure of proteins so as to appreciate and utilise the functionalities of the proteins.

There are four levels of structural organisation of proteins, these are, primary (1°), secondary (2°), tertiary (3°) and quaternary (4°) structures. The primary structure of a protein is concerned with its covalent structure i.e., the sequence in which the amino acids constituting the protein are arranged. The secondary and the higher order structures are related to the three dimensional structure or the conformational aspects.

3.4.1 Covalent or Primary Structure of Proteins

Proteins are made up by linear linkages between different amino acids from a pool of about twenty coded amino acids. These amino acids are present in varying amounts in different proteins and are linked in defined sequences determined by the respective genetic code followed by post-translational modifications (if any). The direct determination of the covalent structure of a given protein is almost an impossible task, as a protein with just 100 residues will have about 1500 atoms whose relative position needs to be determined to ascertain its structure. This complexity, however, is decreased by the fact that proteins are made of defined units viz. amino acids. Different proteins differ in terms of the composition of the amino acids and the sequence in which these are linked together. Therefore, there are two important aspects in the primary structure determination of proteins, one is knowing the total number of amino acid residues present in the protein and second is finding out correct sequence of these amino acids. Let us see how do we arrive at the sequence of amino acids in a protein. We begin with the determination of the amino acid composition of proteins.

3.4.1.1 Amino Acid Composition of Proteins

The amino acid composition of a polypeptide (or protein) is determined by a process called amino acid analysis which involves the following steps.

(a) Hydrolysis of the protein into its constituent amino acids
(b) Chromatographic separation of amino acids
(c) Quantification of each amino acid

(a) Hydrolysis of the Protein into its Constituent Amino Acids

Traditionally, a protein is hydrolysed into its constituent amino acids by heating it with 6 M HCl for 24-72 hrs at 110°C under anaerobic conditions in an incubator. The peptide bonds normally hydrolyse at similar rates however, the ones between the amino acid residues with bulky nonpolar side chains require longer periods.

The hydrolysis is carried out in presence of an acid because with alkali there is a risk of racemisation of the amino acids and some amino acids like, arginine, cystine, serine and threonine etc. are even destroyed. Though acid hydrolysis is a method of choice, some of the amino acids may also get damaged in this process. For example, the indole ring of tryptophan is almost completely destroyed by the chlorine produced from the oxidation of HCl which also affects the tyrosine side chain (though partially). The acid hydrolysis also affects the cysteine residue. The extent of destruction of the amino acids containing OH group in their side chains, viz., Ser, Trp and Thr depends on the time of hydrolysis. These are destroyed to the extent of about 10 % per 24 hrs. In such cases the hydrolysis is carried out for different extents of time (24, 28 and 72 hrs.) and the results are extrapolated to zero time to get the true amino acid content. On the other hand, the peptide bond between valine and isoleucine (if present) is relatively resistant to hydrolysis and may need the hydrolysis to continue for upto ~5 days. The amino acids which are destroyed in the acid hydrolysis are also determined by alternative methods like, alkaline hydrolysis, action of proteolytic enzymes or spectroscopic methods.

Further, in the acid hydrolysis it is not possible to distinguish between glutamine and glutamic acid or between aspargine and aspartic acid. The acid hydrolysis results in the loss of NH_3 from the side chain amide group of glutamine and asparagine, with the consequent production of glutamic and aspartic acid respectively. The estimates of amino acid composition based on acid hydrolysis therefore show glutamine and glutamic acid combined and measured as glutamic acid. Similarly asparagine and aspartic acid are measured together as aspartic acid. The amounts of asparagine and glutamine are estimated from the amount of ammonia liberated during hydrolysis. The hydrolysate so obtained is subjected to chromatographic separation.

(b) Chromatographic Separation of Amino Acids from Protein Hydrolysate

The chromatographic separation of amino acids from protein hydrolysate-the product of hydrolysis, is routinely done with the help of **ion exchange**

chromatography. The separated amino acids are detected and quantified by using ninhydrin or some fluorescent reagent. All these processes are fully automated. In ion exchange chromatography the separation is based on the differential affinity of the amino acids and the resin in the chromatographic column. Sulphonated polystyrene resin having highly polar SO_3H groups are commonly employed in these columns. The amino acids are absorbed by the resin because of attractive forces between the negatively charged sulphonic groups and the positively charged amino acids. The absorption increases with the basicity of the amino acid. Conversely, if a polymer (resin) with positive charge is used then negatively charged amino acid would have a better affinity to the polymer or the resin employed. Thus the forces responsible for differential binding of amino acids to an ion exchange resin depend on:

- side chain polarity
- differences in pK_a values of α-amino and carboxyl groups
- solvent effects etc.

These amino acids are then eluted with the help of an eluting buffer of varying pH. The pH of eluting buffer is changed during the separation to ensure an effective separation of the amino acids in the mixture.

Modern amino acid analysers employ more sensitive and rapid procedures. In these, the amino acid mixture in the protein hydrolysate is reacted with a suitable reagent like, phenylisothiocyanate (PITC) to get their derivatives before subjecting to separation. These derivatives are then subjected to reverse phase chromatography to separate and are then quantified as discussed below.

(c) Quantification of Amino Acids

After eluting from the column, the eluate is allowed to mix with ninhydrin for quantitative determination of the separated amino acids. The reagent reacts with α-amino acids to give an intense purple (λ_{max} 570 nm) coloured product called **Ruhemann's purple**. The observed intensity is a measure of the amount of the amino acid. Proline and hydroxyproline give different products (absorbing at 440 nm) with ninhydrin because their α-amino groups are secondary amines and are a part of a five membered ring.

This operation is also fully automated in an amino acid analyser. The amino acid analyser is first standardised by a mixture containing known quantities of amino acids. Different amino acids separate out and give characteristic peaks at different elution times. The areas under the curve being proportional to the amount of the amino acid. It is followed by running the protein hydrolysate through the amino acid analyser. The column is eluted with same buffer system as used in the standardisation procedure. The peaks obtained are used to identify amino acids and areas under the peaks give an estimate of their relative ratios. These in turn can be converted into their molar ratios or say empirical formula of the peptide. Knowledge of the molar mass of the protein determined by physical (e.g., electrophoresis, ultracentrifugation etc.) and chemical methods can be used to convert these ratios to number of residues of different amino acids per chain. Though these numbers are ought to be integers but due to the errors associated with the determination of the molar mass and the amino acid composition, these turn out to be nonintegral.

The next important step is determination of the sequence i.e., the order in which these amino acids are linked to form a protein.

3.4.1.2 Amino Acid Sequence of Polypeptides

The linear arrangement, of amino acids constituting the proteins in a definite sequence is called the **primary structure** of the proteins. The primary structure defines the protein unambiguously i.e., every protein has a specific amino acid sequence. It is this sequence that determines the chemical, structural and the biological properties of the protein. It is therefore pertinent to determine the sequence of the amino acids or the primary structure of a protein. A large number of sequences is possible for even a relatively small peptide. A tripeptide composed of 3 different amino acids (say A, B and C) can have six different amino acid sequences.

A–B–C,   A–C–B,   B–A–C,   B–C–A,   C–A–B   and   C–B–A

A tetrapeptide with four different amino acids on the other hand can have as many as 24 sequences and for proteins with more than 50 amino acids this number becomes astronomical and only one of these is the actual sequence. Thus, the protein sequence determination is a formidable task. However, a wide range of procedures are now available that may be used in isolation or in a combination there of to decipher the correct amino acid sequence of any known protein. Before proceeding to the sequence determination we need to obtain a pure polypeptide chain free of inter and intramolecular disulphide linkages.

The amino acid sequence determination involves the following steps.

A. Purification or separation of proteins

B. Cleavage of inter and intramolecular disulphide linkage

C. Determination of the terminal amino acid residues

D. Specific cleavage of the polypeptide chain into small fragments by partial hydrolysis

E. Separation and sequence determination of the fragments

F. Determination of the overall sequence from the fragment sequences with appropriate overlap

(A) Purification or Separation of Proteins

The purification of proteins is quite an involved process. A number of methods are available for this purpose, each with its advantages and disadvantages and applicability. Any of these techniques or a combination of these can be used to obtain a purified protein. Some of these methods are:

 (i) **Differential precipitation:** The solubility of protein depends on the concentration of the salt present in the solution (Sec.3.3.4). Different proteins show different dependence of the solubility on the nature and concentration of the salt. This fact is exploited to precipitate (salt out) the desired protein.

 (ii) **Column chromatography:** The column chromatographic technique exploits differential binding of the protein to the column material. Depending on the nature of the column material and the interaction involved with the protein a number of techniques like, ion exchange (proteins have differential affinity for the charged column), affinity chromatography (here the column material has special binding affinity for the protein) etc. are known. Another technique called gel-exclusion chromatography exploits the size differential of different proteins for their separation /purification. Nowadays the modern chromatographic techniques (e.g., hplc) have an added feature of high pressure. The eluent is passed through the column under extremely high pressures. This affords a rapid and excellent separation even at very low concentration by using highly fine columns.

 (iii) **Gel electrophoresis:** This technique is based on the differential electrophoretic mobilities of the proteins which in turn are due to the difference in their isoelectric points. SDS (sodium dodecyl sulphate) gel electrophoresis is the most commonly employed electrophoretic method. In this method SDS –a surfactant, denatures the protein and binds to it. The number of molecules binding to the protein depends on the size (molecular weight) of the protein; a high molecular weight protein will bind more molecules of the surfactant. Further, since each molecule of the bound SDS contributes a negative charge to the protein, different proteins acquire different charges depending on their size. These are accordingly separated in presence of electric field on the basis of the charges on them.

(iv) Differential centrifugation: In this method the protein mixture is centrifuged at different speeds and for different times. This leads to the separation of the protein molecules - each coming at different speed and time. The heavier proteins would need lower speeds and lesser time while the lighter (smaller) ones would separate at higher speeds and require more time.

Polypeptides or proteins purified by **differential precipitation** or **column chromatographic methods** are subjected to sulphide cleavage by oxidising the disulphide linkage with performic acid or reducing it with thioalcohol as discussed below.

(B) Cleavage of Inter and Intramolecular Disulphide Linkages

The cysteine residues present in the primary structure of the protein are linked through inter and intramolecular disulphide linkages. It means that the cysteine residues within a given protein molecule or in two different molecules are joined together. For example, in the protein hormone insulin there are two polypeptide chains (called chain A and chain B) which are linked together through two disulphide bonds while a disulphide bond exists between the two cysteine residues within chain A. Therefore, before attempting sequence determination these disulphide bonds should be cleaved. This is accomplished by either oxidising the linkage with performic acid or by its reduction with thiol reagents (containing SH groups) such as mercaptoethanol.

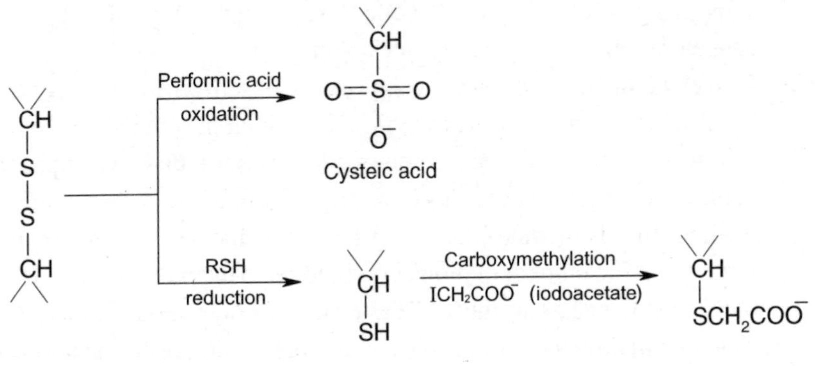

The cleavage of the disulphide bonds disengages the interlinked protein chains (if any) which are then separated by the separation / purification methods discussed above. Once purified, these proteins are then subjected to sequence determination.

(C) Determination of the Terminal Amino Acid Residues

In the first step of actual sequence determination the end groups or the terminal residues are identified. A number of methods are available for the purpose and require that the N-terminal amino acid has a free amino group and the C-terminal amino acid has a free carboxyl group. Let us begin with the determination of the N-terminal residue.

(a) Determination of the N-Terminal Residue

The N-terminal residue can be identified by derivatising all the amino groups (the derivatives must be stable to hydrolysis) present in the protein followed by hydrolysis of the protein. The hydrolysate so obtained contains only one amino acid that is labelled / derivatised at the amino terminal. Other amino acids will have their derivatisation (if any) in the side chain only e.g., in case of lysyl residue the ε-amino group would be derivatised. The derivatised N-terminal amino acid can be identified by TLC or paper chromatography. The two most commonly used derivatising agents are fluoro-2, 4-dinitrobenzene (FDNB) and dansyl chloride. The N-terminal determination methods based on these, along with a few more agents are discussed below.

(i) **Sanger or FDNB method:** This method was developed by Frederick Sanger who used it in the amino acid sequence determination of insulin. He was awarded the Nobel prize in Chemistry in 1958 for this work. In this method the polypeptide on heating with fluorodinitrobenzene (FDNB) in mildly basic solution, undergoes a nucleophilic aromatic substitution (S_NAr) to give a N-terminal derivative of the protein. Hydrolysis of the derivatised protein gives a mixture of amino acids in which the N-terminal amino acid is labelled with 2,4-dinitrophenyl group (DNP) which is yellow coloured.

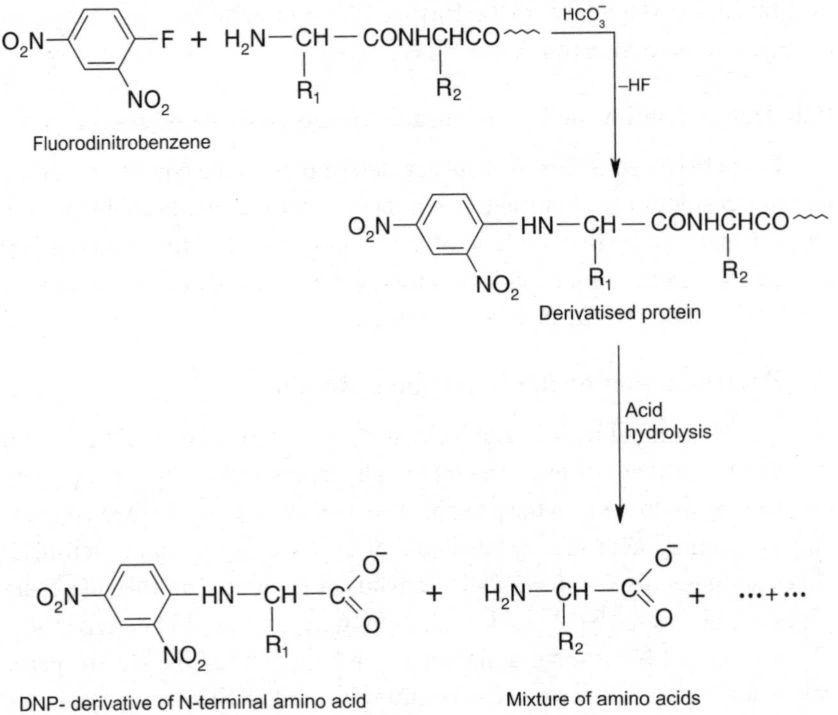

Fluorodinitrobenzene

Derivatised protein

Acid
hydrolysis

DNP- derivative of N-terminal amino acid   Mixture of amino acids

The DNP derivatised N-terminal amino acid can be identified by its characteristic migration rate on thin layer chromatography or paper electrophoresis.

(ii) **Dansyl method:** This is a much more sensitive method than the Sanger's method. It involves the reaction of the terminal amino group of the protein with dansyl chloride (5-dimethylaminonapnthalene-1-sulphonylchloride) to give N-dansyl derivative. The derivatised protein is then hydrolysed to give a highly fluorescent N-dansyl amino acid and a mixture of unprotected amino acids.

The dansylated amino acid is detected chromatographically or with the help of its characteristic fluorescence in the UV region.

The reaction can be represented as follows.

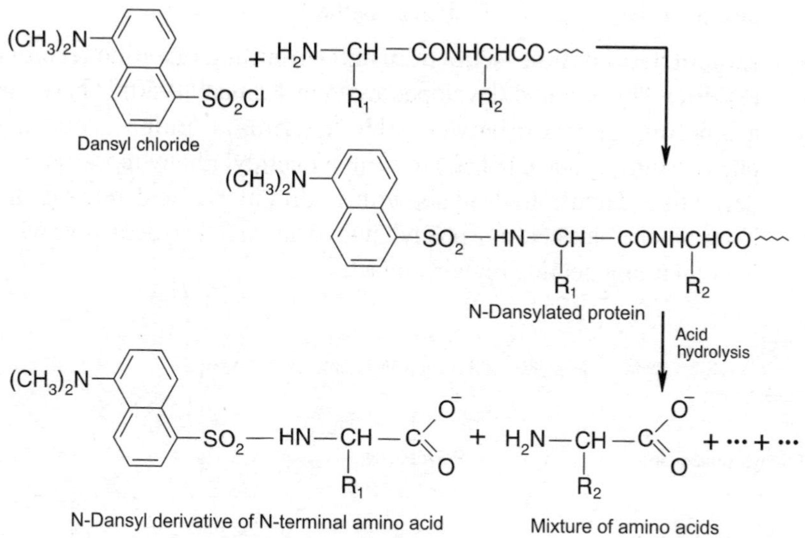

Dansyl chloride

N-Dansylated protein

Acid hydrolysis

N-Dansyl derivative of N-terminal amino acid

Mixture of amino acids

(iii) **Cyanate method:** In this method, the free amino group of the peptide is reacted with cyanate. Consequent acid hydrolysis of the derivatised protein converts the labelled residue to a hydantoin while the other amino acids are normal. The identification of the hydantoin reveals the identity of the terminal amino acid.

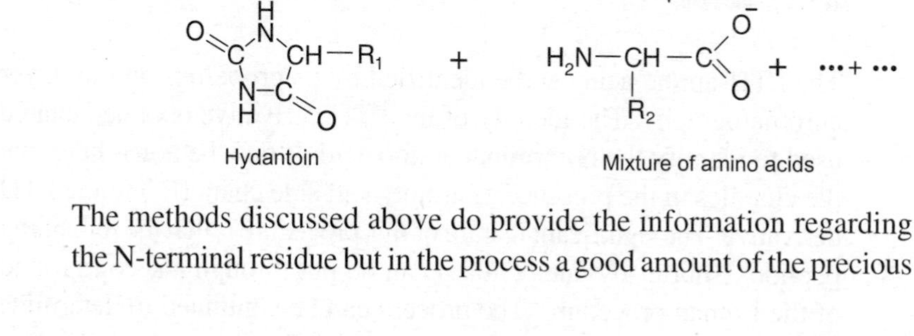

Acid Hydrolysis

Hydantoin

Mixture of amino acids

The methods discussed above do provide the information regarding the N-terminal residue but in the process a good amount of the precious

protein is lost. It is desirable to be able to determine the terminal
residue without destroying the rest of the protein. Edman degradation
discussed below provides such a method.

**(iv) Edman degradation– sequential determination of amino terminal
residue:** This method developed by Pehr Edman (1950) is based on
a labelling reaction between the N-terminal amino group and
phenylisothiocyanate (PITC) to form a peptidyl phenythiocarbamoyl
derivative. Gentle hydrolysis with hydrochloric acid releases the
N-terminal amino acid as a phenylthiohydantoin (PTH) derivative while
the remaining peptide remains intact.

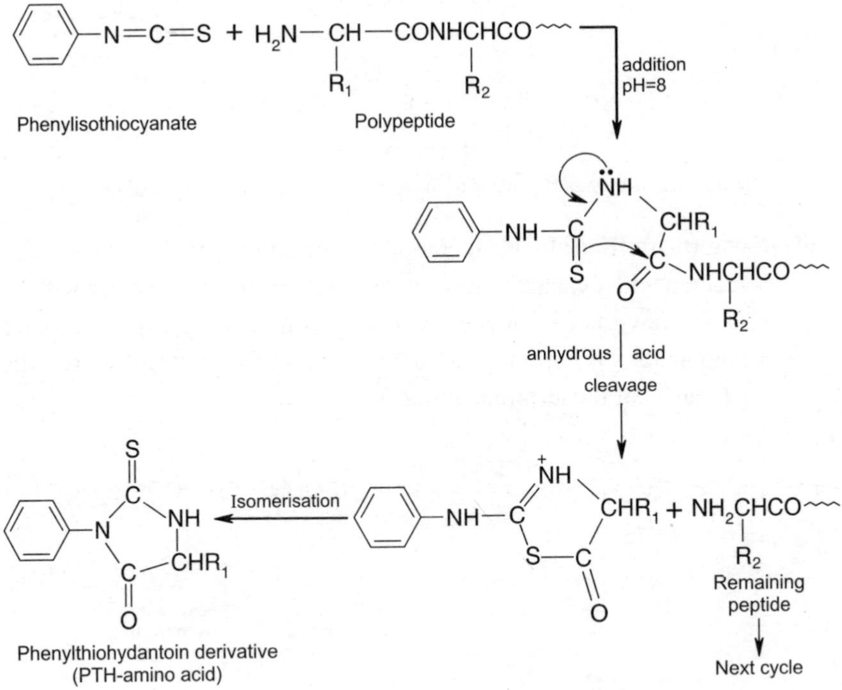

Phenylthiohydantoin derivative
(PTH-amino acid)

The PTH-amino acid can be identified by its properties on thin layer
chromatography. The identity of the PTH derivative obtained can be
used to identify the N-terminal amino acid. It may be noted here that
the clue lies in the presence of amino acid side chain (R_1) in the PTH
derivative. The significant feature of this procedure is that the remaining
peptide (shorter by one residue) can be put through a second round
of the Edman procedure. The process can be continued to determine
the whole sequence one by one. In fact, there is a device called

sequenator which automates the Edman degradation procedure and each amino acid is automatically detected as it is removed.

(b) Determination of the C-Terminal Residue

There are not many satisfactory methods available for the determination of the C-terminal residue of a polypeptide or a protein. Some of the better known methods are as follows.

(i) **Hydrazinolysis method:** The most commonly used method for the determination of C- terminal residue is based on hydrazinolysis reaction. It involves the treatment of the polypeptide with anhydrous **hydrazine** at 100°C. This results in the conversion of all amino acid residues, except the C-terminal residue, to the corresponding hydrazides. The C-terminal amino acid which remains as the free amino acid can in principle be isolated and identified chromatographically, though it is not easy. On passing the mixture over a column of strong cation exchange resin and eluting, the free amino acid comes out while the basic hydrazides are retained on the column.

$$
\underset{\substack{\\ R_1 \; \text{Polypeptide} \quad R_n}}{H_2N-CH-\overset{\overset{\displaystyle O}{\|}}{C} \cdots\cdots HN-CH-COOH} \xrightarrow{\;H_2N-NH_2\;}
$$

$$
\underset{\substack{R_n \\ \text{C-terminal amino acid}}}{H_2N-CH-COOH} + \underset{\substack{R_1 \\ \text{Hydrazides}}}{H_2N-CH-CONHNH_2} + \cdots
$$

Further, if asparagine or glutamine happen to be the C- terminal residues these are not recovered as free amino acids. The side chain amide group of these amino acids are converted into hydrazides while the amino acid arginine becomes ornithine- an amino acid similar to lysine with one CH_2 group less in the side chain.

(ii) **Enzymatic method:** In this method, the polypeptide is treated with a proteolytic enzyme *carboxypeptidase* that results in a limited breakdown of the polypeptide. The enzyme (obtained from pancreas) cleaves only peptide linkages adjacent to free α-carboxyl groups. The free amino acid so released can then be identified chromatographically after suitable labelling with ninhydrin or other agents used in amino acid analysis. The process can be used repeatedly to get the new C-terminal residue identified in a stepwise manner. However, these carboxypeptidases show target specificities. For example, *carboxypeptidase A* does not cleave Arg or Lys residues and also the residue next to proline. *Carboxypeptidase B* hydrolyses only Arg and Lys and that too if not preceded by proline. Therefore, we may need to employ a number of carboxypeptidases to determine the sequence of the peptide by this method.

(iii) **Thiocyanate method:** In addition to the enzymatic method a chemical procedure involving thiocyanate is also available for sequencing of polypeptide from the C-terminal. This method is not as successful as the Edman's procedure for N-terminal sequence determination. In this method the treatment of the free carboxyl group of the polypeptide with acetic anhydride gives a mixed anhydride. Action of thiocyanate on this gives a new mixed anhydride which gives a hydantoin through cyclisation. The C- terminal amino acid cleaves as a thiohydantoin on hydrolysis and is analysed. The remaining polypeptide can be put through the next round of procedure and so on.

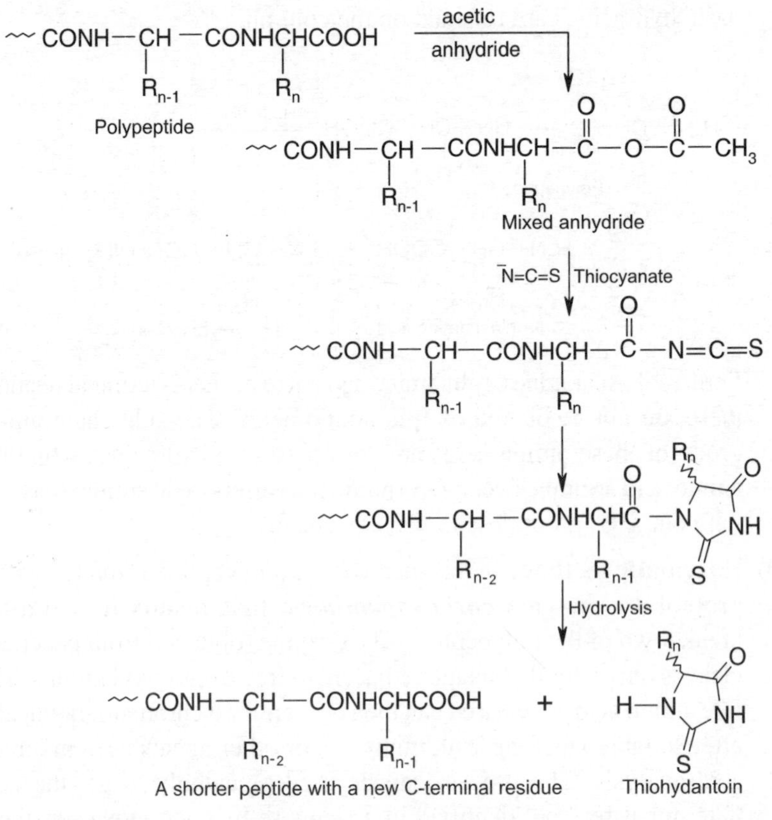

(iv) **Treatment with LiBH$_4$:** In yet another procedure, the polypeptide is treated with $LiBH_4$ followed by acid hydrolysis. Here the carboxyl group gets reduced to alcohol which can be analysed to identify the amino acid.

The sequential determination of the primary structure from N-terminal (to some extent C- terminal) is dependable for short sequences only. There are problems with long sequences. The biggest problem is associated with the non-quantitative liberation of the terminal residue. If a significant amount of a residue remains unliberated in a given cycle it would appear in the next cycle and interfere with the determination of the next residue. The error cumulates over the cycles and leads to a mixture of amino acid residues in the subsequent cycles. To circumvent this problem the polypeptide chain is partially hydrolysed to get a mixture of short peptides which are then subjected to sequence determination. This partial hydrolysis is done in more than one way so as to get different sets of peptides. The results of the sequence determination of the fragments are then put together to reconstruct the overall sequence.

(D) Specific Cleavage of the Polypeptide Chain into Small Fragments by Partial Hydrolysis

The terminal analysis done by all the above methods is useful for short chain polypeptides of say upto about fifty amino acid residues. For larger proteins it is necessary to cleave them to smaller peptides. The chains can be broken to smaller fragments by using dilute acid or enzymes for partial hydrolysis. The significance of partial hydrolysis can be understood by taking a simple example. Suppose we have a tripeptide containing three different amino acids A, B and C. There are six ways (p-134) in which the three amino acids could be arranged to form the tripeptide. Suppose the partial acid-catalysed hydrolysis of this tripeptide yields two dipeptides, A-C and C-B. The hydrolysis makes clear that C is the middle amino acid and the correct sequence of three amino acids is A-C-B. Different chemical and enzymatic methods are available for fragmentation of protein into smaller peptides. Some of the commonly used are given below:

(i) **Cleavage with proteolytic enzymes:** Certain enzymes cause the hydrolysis of polypeptides at specific points. These are called proteases or proteolytic enzymes. A number of such enzymes are available and are routinely employed for the purpose. Depending upon whether the enzyme cleaves the peptide at internal or terminal residues it is called an ***endopeptidase*** or an ***exopeptidase*** respectively. Trypsin probably is the most specific enzyme. It cleaves the peptide bond next to the basic residues like lysine and arginine in the polypeptide. The specificity can be further increased to arginine only by derivatising the ε- amino group of the lysyl residues. Alternatively the applicability of trypsin can be extended to cysteine by derivatising its side chain thiol group with ethyleneimine whereby it becomes a basic residue. Chymotrypsin, another important endopeptidase cleaves the polypeptide chain at the peptide bond following the aromatic residues viz., tyrosine, tryptophan and phenylalanine. It is not as specific as trypsin as it also cleaves the peptide bond following the hydrophobic amino acid residues like leucine. Pepsin on the other hand cleaves the peptide bond preceding aromaic residues like phenylalanine and tyrosine. The cleavage sites (marked by an arrow in the following structure) of some of the common proteolytic enzymes are given in Table 3.4.

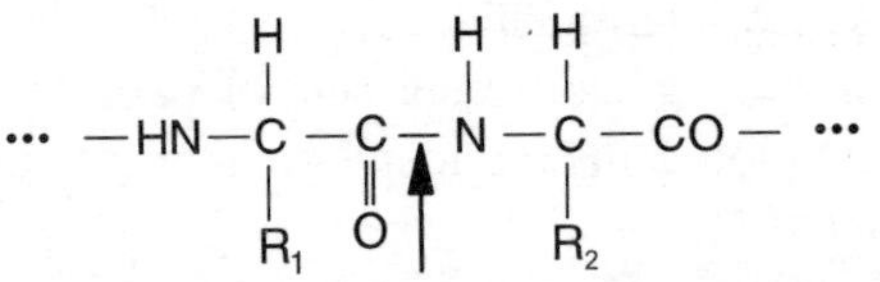

$$\cdots -HN-\underset{\underset{R_1}{|}}{\overset{\overset{H}{|}}{C}}-\underset{O}{\overset{}{\underset{\|}{C}}}\uparrow N-\underset{\underset{R_2}{|}}{\overset{\overset{H}{|}}{C}}-CO-\cdots$$

Table 3.4   Cleavage sites of some of the common proteolytic enzymes

Proteolytic enzyme	Cleavage site
Trypsin	R_1 = Lys or Arg; $R_2 \neq$ Pro
Pepsin	R_2 = Tyr, Phe
Carbopeptidase A	$R_1 \neq$ Pro; R_2 = C-terminal amino acid
Chymotrypsin	R_1 = Tyr, Phe, Trp, Leu, Ile or Val; $R_2 \neq$ Pro
Thermolysin	$R_1 \neq$ Pro; R_2 = Ile, Met, Trp, Tyr, Phe and Val

Insulin was the first polypeptide to have its complete amino acid sequence determined by Sanger during 1951-1955. Since that time the structures of many other polypeptides and proteins have been elucidated.

(ii) **Cleavage at methionine with cyanogen bromide:** The peptide when treated with cyanogen bromide cleaves specifically at the carboxyl side of the methionine residue. Cyanogen bromide acts like an acid halide. Though it can react with any nucleophilic amino acid side chain, under the above reaction conditions, it reacts only at methionine to result into a cleavage. As far as the mechanism of the reaction is concerned, it can be shown as given below. The sulphur in the methionine side chain acts as a nucleophile, displacing bromide from cyanogen bromide to give a type of sulphonium ion.

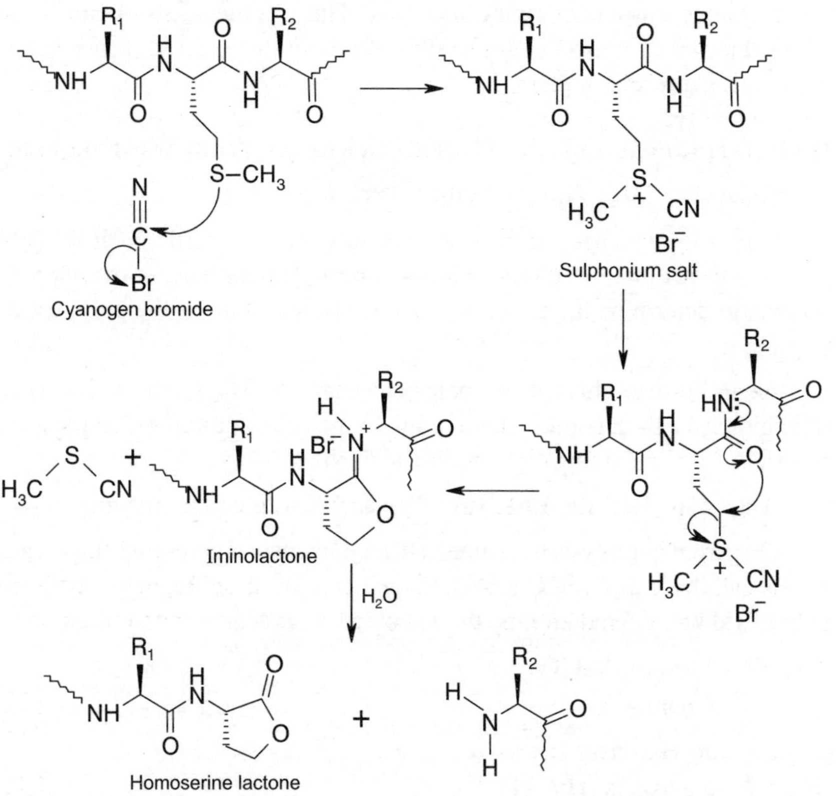

Cyanogen bromide

Sulphonium salt

Iminolactone

Homoserine lactone

The stereochemistry of the methionine side chain is favourable for an intramolecular rearrangement of the sulphonium ion. The sulphonium ion – an excellent leaving group with its electron withdrawing cyanide, is displaced by the oxygen of the neighbouring amide bond to form a five membered ring.

The last step involves the hydrolysis of the iminolactone by water to cleave the peptide bond in the polypeptide chain. The original methionine residue now exists as a homoserine lactone which hydrolyses to give free acid.

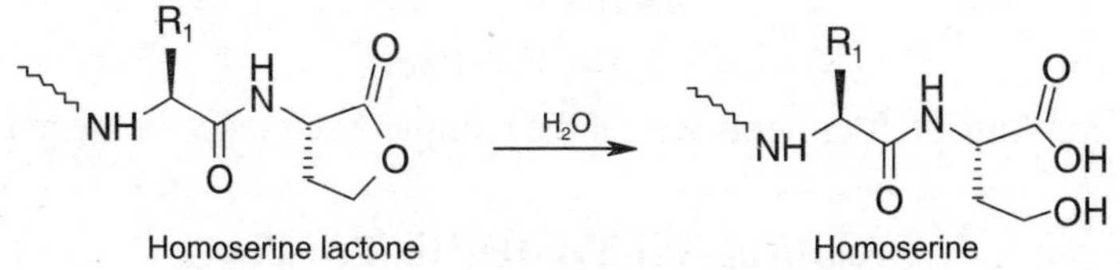

Homoserine lactone   Homoserine

(E) Separation and Sequence Determination of the Fragments

Peptides resulting from the cleavage of polypeptides are separated by ion-exchange chromatographic methods. The fragments so obtained are then subjected to Edman degradation to find out the exact sequence of the amino acids present in them.

(F) Determination of the Overall Sequence from the Fragment Sequences with Appropriate Overlap

Once the sequences of the fragments obtained by partial hydrolysis of the protein are obtained, these are then connected on the basis of overlapping regions to determine the primary sequence or structure of the polypeptide or protein.

Let us illustrate this with the help of an example. The amino acid analysis of angiotensin II– a peptide involved with the regulation of blood pressure shows it to be containing eight amino acids viz.,

Arg, Asp, His, Ile, Phe, Pro, Tyr and **Val** in equal amounts.

On partial hydrolysis with dilute HCl angiotensin II gives four fragments. Let us call these as A, B, C and D. Sequencing of these fragments with the help of Edman degradation method showed their sequences to be as:

A :   Asp-Arg-Val-Tyr

B :   Pro-Phe

C :   Ile-His-Pro

D :   Val-Tyr-Ile-His

A look at the sequences of fragments A and D shows that the sequence Val-Tyr is common in them. Writing these fragments with suitable alignment of the overlapping residues, we get,

             Asp-Arg-Val-Tyr

                    Val-Tyr-Ile-His

Continuing in the same fashion for the fragments C and B we get,

             Asp-Arg-Val-Tyr

                    Val-Tyr-Ile-His

                           Ile-His-Pro

                                  Pro-Phe

Assembling the fragments we get the complete sequence of the peptide to be,

Asp-Arg-Val-Tyr-Ile-His-Pro-Phe

3.4.2 Conformational Aspects of Proteins: Higher Order Structures

Proteins are large molecules in which the rotation around different bonds can give rise to various three dimensional arrangements of atoms. These arrangements are interconvertible without breaking the covalent bonds and are referred to as the **conformations**. Due to the size and possible flexibility of proteins innumerable conformations are possible. However, due to several factors the structural options actually observed are relatively fewer. The properties of peptides and proteins depend not only on their component amino acids and their sequence in peptide chains, but also on the way in which the peptide chains are stretched, coiled and folded in space. The structure adopted by proteins can be understood in terms of structural organisation at different levels. In the hierarchy of protein structural organisation the lowest level i.e., the primary structure refers to the linear sequence of amino acids in a polypeptide chain held together by peptide bonds. As a part of an amide functional group, these bonds are difficult to break, so the sequence of a protein is quite stable. This along with its determination has already been discussed in detail (Sec. 3.4.1). The secondary structure refers to regular local structures like, a α-helix or a β-sheet etc adopted by linear segments of polypeptide chains. Tertiary structure is concerned with the overall topology of the polypeptide or protein resulting from long-range contacts within the chain while the quaternary structure is the organisation of two or more independent polypeptide chains. Let us first take up the secondary structure of the proteins.

3.4.2.1 Secondary Structure of Proteins

According to IUPAC-IUB (1970) **'the secondary structure of a sequence of a segment of polypeptide chain is the local spatial arrangement of its main chain atoms without regard to the conformation of its side chains or its relationship with other segments'.** Thus secondary structure is basically the specific geometric arrangement of the amino acids that results from peptide linkages that are close to each other in the polypeptide chain of a protein. An extended polypeptide chain has a regularly repeating part called the main chain (or backbone) consisting of alternating peptide bonds and α-carbon atoms and a variable part comprising of the distinctive side chains. Let us look at a portion of the polypeptide backbone given in Fig 3.2 to understand the conventions used in describing polypeptide conformations.

Chemistry of Natural Products

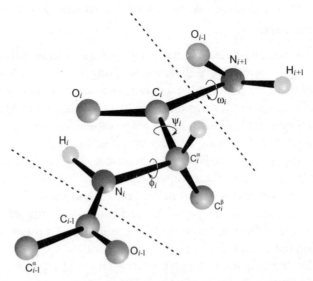

Fig. 3.2 *A portion of the polypeptide backbone in fully extended conformation.*

The peptide bonds as discussed earlier (chapter 2, Fig.2.1) are rigid planar units. As shown in Fig. 3.2, two peptide units are attached to a α-carbon atom marked as C_i^α. The subscripts 'i' refer to the number of the amino acid residues in the polypeptide chain. The rotations around different bonds are described in terms of **dihedral** or **torsion angles**. The rotation about the N–C^α bond of the backbone is denoted by torsion angle φ while the rotation around C^α–C_i is denoted by the torsion angle ψ. In the maximally extended chain (as shown in the figure) i.e., when N_i C^α and C_i atoms are all *trans* to each other the value of φ and ψ is given as 180° (or –180°). The rotation around the peptide C–N bond is expressed in terms of angle ω. The rotation around these bonds are given positive values when the atom behind the bond moves in the clockwise direction and the movement in the anticlockwise directions is indicated by negative values.

Though the rotation about the N_i–C^α and C^α–C_i single bonds is free, the possible φ and ψ values for an amino acid residue in a peptide are geometrically constrained due to possible steric clashes between the neighbouring atoms. G.N. Ramachandran and colleagues determined these torsional angles using hard sphere models and represented them in terms of a two dimensional plot called as Ramachandran plot. Fig.3.3 shows the Ramachandran plots for glycine and alanine. The shaded region in the plot indicates the allowed (i.e., no steric overlap) values for the torsion angles while the solid lines indicate the highly unfavourable values. The region enclosed by these solid lines is partially allowed. The dashed connecting lines enclose the regions that may be allowed if the bond angles are slightly altered. An observation of the Ramachandran plots for glycine and alanine reveal that glycyl residue shows much greater flexibility as compared to alanine. This is due to the absence of a β- carbon atom in the side chain. In glycine 45% of the total φ ψ space is fully allowed while for alanine it is merely 7.5%. In the extreme limits these numbers are 61% and 22.5% respectively.

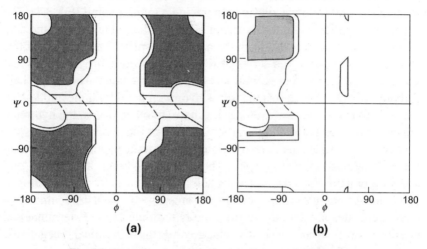

Fig. 3.3 *Ramachandran plots for (a) glycine and (b) alanine.*

It is due to these restrictions in rotation and the rigidity of the peptide bonds that the possible conformations of a protein molecule are limited. On the basis of extensive work on the X-ray diffraction studies of synthetic polypeptides and model building, Pauling and Corey identified two ordered arrangements of the polypeptide backbone stabilised by extensive hydrogen bonding. These particularly stable ordered arrangements of proteins are the **α-helix** and the **β-sheet** structures. Hydrogen bonding and van der Waals forces in or around these arrangements make the backbone chain relatively stiff and resistant to bending. In addition **reverse turns** (or bends) form another standard secondary structure. A less regular arrangement called the **random coil** provides flexible regions allowing amino acid chains to bend and fold. The particular secondary structure adopted by a given protein is completely determined by its sequence. Factors affecting the secondary structures of peptide chains are:

- the planarity of peptide bonds
- hydrogen bonding of amide carbonyl groups to NH groups
- the hydrophilic and hydrophobic character of substituent groups
- steric crowding of neighbouring groups
- attraction and repulsion of charged groups

Different secondary structures possible for proteins are discussed below.

α-Helix

The α-helix first proposed by the nobel laureate Linus Pauling and Robert Corey is the best known and recognised secondary structure of the proteins. In this conformation, the alternating α-carbons and peptide linkages in the backbone follow a zigzag path tracing a regular spiral (Fig.3.4). The coiled polypeptide chain forms the inner part of the rod and the amino acid side chains extend outward from the axis of the spiral. The stability of this structure is attributed to a series of regularly spaced hydrogen bonds and van der Waals forces between atoms in the main chains. The main chain carbonyl (C=O) of each residue forms hydrogen bond with the amide NH, which is four residues away along the chain. In other words, the hydrogen bonds are formed between the carbonyl group of the peptide bond existing between residue n and n+1 and the amino group of the peptide bond existing between the residues n+3 and n+4. These hydrogen bonds are 286 pm long and run along the chain parallel to the helix axis. In the helix all the hydrogen bonds and the peptide groups point in the same direction and their cumulative effect generates a macrodipole from the N-terminal to the C-terminal end i.e., the N-terminal bears a positive charge while the C- terminal is negatively charged, Fig.3.4 (b).

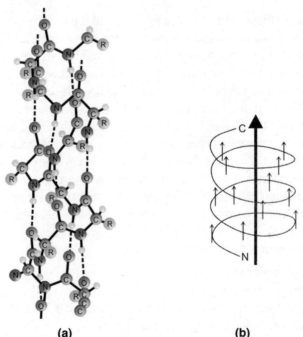

(a)   (b)

Fig. 3.4 *Secondary protein structure: (a) α-Helix (b) Macrodipole in a helix.*

Although an alpha helix may turn in either direction, the alpha helices of almost all natural proteins are right handed which means that it rotates clockwise as it spirals away from a viewer at either end. In simple words, if we start at any point and move clockwise along the backbone of the polypeptide we move forward. In a left handed helix we need to move anticlockwise to move forward.

Each residue is related to the next one by a translation of 15 pm and a rotation of 100° along the helix axis, which gives 3.6 amino acid residues (360°/100°) per turn of the helix. Thus amino acids which are three to four residues away in the linear sequence are spatially quite close to one another in an α-helix. The φ and ψ values for the α-helix are −57° and −47° respectively. The relative locations of the donor and acceptor atoms of the hydrogen bond, the number of amino acid units per helical turn and pitch (the linear distance travelled along the helical axis per turn) are the structural features that define α-helix. What we have just described is a classical helix. In actual proteins the detailed geometry of the helices is different from the classical helix. The standard or the classical helix is just one of the possibilities.

α- Helices in which the chain is either more tightly or relatively loosely coiled are also known. In these the hydrogen bonding takes place between the peptide bond of a given residue (i^{th}) and the one three residues (i+3rd) and five residues (i + 5th) away. These are designated as 3_{10} helix and π-helix respectively.

These helices are usually found at the end of regular helices, 3_{10} helices being more common than the π helices. In these helices the dipoles are not aligned as in the standard helix. In general, helical conformations of peptide chains are described by a two number term, $\mathbf{n_m}$, where $\mathbf{n}$ is the number of amino acid units per turn and $\mathbf{m}$ is the number of atoms in the smallest ring defined by the hydrogen bond. Using this terminology, the standard α-helix is a 3.6_{13} helix which means that this helix has 3.6 residues per turn and there are 13 atoms between hydrogen bond donor and acceptor along the backbone. The φ and ψ values and other helical parameters of different types of α-helices are summarised in Table 3.5.

Table 3.5 Dihedral angles and helical parameters of different types of α-helices

Structure	Φ (deg.)	ψ (deg.)	n (residues/turn)	m	Pitch p (pm)	H–bonding residues
Standard helix	−57	−47	3.6	13	54	i and i+2
3_{10} helix	−74	−4	3.0	10	60	i and i+3
π helix	−57	−70	4.4	16	50	i and i+5

Different amino acids have different tendencies for forming α-helices; glutamic acid, alanine, arginine and lysine prefer α-helix while asparagine, tyrosine and glycine are quite unlikely to form them. In case of the amino acid proline there is a restriction on the rotation around the α-carbon atom, therefore it cannot be accommodated in the α-helical organisation and in fact, it acts as a **helix breaker**.

Theoretically speaking a left handed α-helix is also possible. However in such a conformation the side chains are very close to the backbone, which makes this conformation far less stable than the right handed version and is rarely encountered in natural systems.

The amino acid sequence of a helical segment of a protein is represented in terms of what is called a **'helical wheel'**. In a helical wheel the amino acids are plotted in a circle at an interval of 100°. A helical wheel is essentially a projection of the amino acid side chains down the helix axis. The helical wheel representation of first twenty amino acids of bee venom toxin, melittin is shown in Fig 3.5.

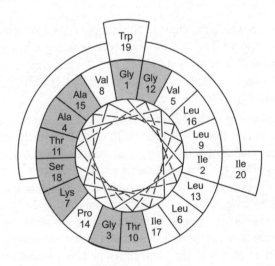

Fig. 3.5 *Helical wheel representation of first twenty amino acids of bee venom toxin, melittin. The hydrophilic residues are shaded.*

In such a representation, the nature of amino acids coming together on a given side of the helix can be used to predict the location of the helix in the overall protein structure. For example, if the nature of amino acids on the either side of the helix is found to be nonpolar then the helix is likely to be buried inside the overall structure. If on the other hand the nonpolar side chains are clustered towards one side and the polar residues on the other side of the helix then it is called **amphipathic** or **amphiphilic helix**. Such a helix is likely to be found on the surface of the protein, the polar side facing the solvent and the nonpolar side facing the interior of the protein. The nalure or helix can also provide clues to its role in the function of protein.

β- Pleated Sheets

Pauling and Corey proposed another regular arrangement of amino acid chains, which is called the **β-sheet**. The basic unit of this structure is a **β-strand** in which the polypeptide backbone chain is fully extended. A β-strand can be visualised as a special helix with two residues per turn (n =2) and having a translation of 320-340 pm per residue. The appropriate backbone dihedral angles are -120° and 120° respectively. In this extended structure the rigid planar peptide bonds force the side chain (R) groups to alternate with no likely interactions between side chains. This otherwise unstable conformation stabilises itself by forming a sheet like structure called β-sheet. The stability of the β-sheet can be attributed to the hydrogen bonds between the -NH and C = O groups of a beta strand with adjacent beta strands aligned side-by-side. The hydrogen bonds in the β-sheets are about 1pm shorter than those in the α-helices.

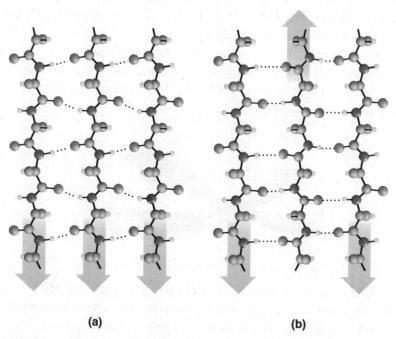

(a)   (b)

Fig. 3.6 *(a) Parallel and (b) antiparallel β-sheet structures. The arrows indicate the N- terminal to C-terminal direction.*

The classical β-sheet structure originally proposed was planar and flat. However in such a conformation the presence of bulky side chains on adjacent β-strands would cause a kind of crowding that make it unstable. The strain is relieved by a right handed twist of about 0-30° between strands. This leads to what is known as a **β-pleated sheet** in which the planes of alternate peptide groups are at an angle to each other and the α-carbons carrying the side chains lie at the line joining these planes. The neighbouring C_β atoms are alternatively above and below the plane of the sheet. The name pleated sheet comes from the fan like structure akin to the hand fan made by pleating a sheet of thick paper. Further, the polypeptide chains in a sheet may run in opposite directions (antiparallel) or in the same direction (parallel).

In a parallel β-pleated sheet structure the N-terminals of different strands are aligned (head to head) while in the antiparallel β-sheet the N-terminal of one strand is aligned with the C-terminal (head to tail) of the other strand and so on. The antiparallel sheets are more twisted than the parallel ones.

The β-sheets can be either of purely parallel or antiparallel variety or a mixture of the two. However, in nature antiparallel sheets are much more preferred as compared to the parallel β-sheets. Since bulky side-chain substituents destabilise this arrangement due to steric crowding, beta-sheet conformation is usually limited to peptides having a large amount of glycine and alanine. Silk fibroin, the fibre from the cocoon of silk moth (*bombyx mori*) is one of the common examples of proteins containing almost entirely antiparallel β-pleated sheets.

Fig. 3.7 *A schematic representation of twisted β-sheets.*

Purely parallel or antiparallel sheets of 6 or 8 strands often turn around and form a closed structure called β-**barrels**. An interesting example of β-barrel was initially found in the enzyme *triosephosphateisomerase* (Fig.3.8) called as TIM barrel. In this structure 8 parallel β-strands form a barrel which is surrounded by an equal number of α-helices connecting each pair of β- strands from outside. Today, at least 16 proteins are known to have this type of structure.

Fig. 3.8 *Triosephosphate isomerase: A β- barrel.*

Reverse Turns

Different segments of a polypeptide chain acquire different secondary structures, discussed above, depending on the nature of the side chains of the residues constituting them. However, to attain the overall globular structure, the polypeptide chain needs to take turns so as to change direction. In fact, nearly one-fourth to about one-third of the residues of globular proteins are involved in turns or loops. The **reverse turns** were initially predicted theoretically by Venkatchalam on the basis of his exploration of possible conformations for a tetrapeptide containing three peptide bonds. He, in fact proposed three types of 'turns' containing hydrogen bond between the carbonyl oxygen of the residue i and the nitrogen of residue i+3. Several definable turns and bends in protein structure have been recognised since then. There are two main types of reverse turns (Fig 3.9) called **β-turns** and **γ-turns** depending on the type of secondary structures they link and the number of residues involved in the process. Commonly observed β-turns refer to the turns found at the place where the polypeptide chain (β-strand) turns to form another β-strand lying besides the first one. Since these turns are located primarily on the surface of the protein, these contain mainly polar and charged residues.

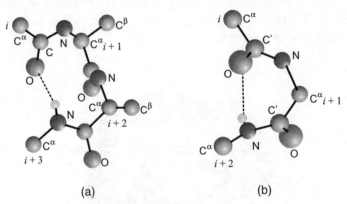

(a)   (b)

Fig. 3.9 *Reverse turns   (a) β -turn   (b) γ - turn*

In this process two residues of the chain are not involved in the hydrogen bonding scheme of the β-sheet. The β-turn is defined in terms of four residues, the two non-hydrogen bonding residues and a residue each on the either side of these. These are designated as i to i+3, the carbonyl group of residue i gets into hydrogen bonding with the N-H of i +3rd residue. The **γ-turns** are the tightest turns involving only 3 residues with hydrogen bonding between the carbonyl group of the 1st residue with the N-H of the 3rd residue. There are variants of these turns are also known but, are beyond the scope of this text.

At the simplest level, proteins can be classified by their content of secondary structures like α-helices and β-sheets. For example, *E. coli* cytochrome B_{562} is composed mainly of α-helices while the green algae *plastocyanin* mostly contains β-sheets. However, many proteins are a mix of α- helices and β-sheets. In addition, there are some proteins where turns and disulfide bonds seem to be structurally more important e.g. in agglutinin, the wheat germ protein, turns and disulfide bonds are the predominant structural components.

Random Coil

In the **random coil** structure, as the name suggests, the amino acid chain takes up an irregular configuration with no tendency to form alpha helices or beta strands or any other distinct pattern. In this structure the conformation of each part of the polypeptide chain is assumed to be independent of the conformation of the rest of the molecule. Proline residues often contribute to random coil structures because their ring form does not fit into an alpha helix or a beta sheet and no sites are available for formation of stabilising hydrogen bonds. The term *random coil* is somewhat a misnomer, because an irregularly bent or folded region in an amino acid chain is usually not completely random in form. Instead, its amino acid sequence and the conformations of nearby segments limit its conformation to certain possibilities. Similarly, the random coil is not necessarily coiled; its conformation may include bends or irregular shapes with no coiling. Segments of random coils are as significant to protein structure as are more regular conformations because they provide opportunities for the amino acid chain to bend back on itself, thereby allowing proteins that contain alpha-helical or beta-sheets to fold into compact, globular forms.

3.4.2.2 Tertiary Structure of Proteins

At a given pH and temperature the amino acid sequence of a protein determines the distribution of different secondary structures like alpha-helix, beta-strand (or sheet) and random-coil segments, discussed above. These secondary structures formed by different regions of polypeptide backbone are linked by turns and loops to acquire an over all three dimensional structure called as the tertiary structure. According to IUPAC-IUB **'the tertiary structure of a protein molecule, or a subunit of a protein molecule is the arrangement of all its atoms in space, without regard to its relationship with neighbouring molecules or subunits'.** In this structure the tightly folded structure have their polar groups on the surface and non-polar groups buried inside. The stability of the tertiary structure is maintained by hydrogen bonds, attraction between positively and negatively charged side groups, van der Waals forces, and polar and nonpolar associations. The net effect of these factors is to establish a distinct three dimensional shape for each protein of a unique sequence. The number and position of disulfide linkage between cystine residues, the location of proline residues and the position at which other substances, such as metallic ion, lipids, and carbohydrates may bind, is also determined by the amino acid sequence. There are a number of ways to represent the folded structure of a protein and the arrangement of secondary structure elements within the tertiary structure. While these simplifications don't show the side chain and main chain interactions that hold the structures together, they do reveal the overall folding pattern. A diagramatic representation of the tertiary structure of a protein, lysozyme is shown in Fig.3.10. Here arrows indicate the β-sheets while the coils represent the helices.

Fig 3.10 *The tertiary structure of enzyme, lysozyme.*

3.4.2.3 Quaternary Structure of Proteins

Many proteins exist naturally as aggregates of two or more polypeptide chains. Each of these polypeptides, called as subunit, is usually folded into an apparently independent globular conformation. These subunits also called **protomers** are usually designated by letters e.g., normal haemoglobin is designated as $\alpha_2\beta_2$ indicating it to be a tetramer of two alpha and two beta subunits. These aggregated structures are referred to as quaternary structures of the proteins. According to IUPAC-IUB **'the quaternary structure of a protein molecule is the arrangement of its subunits in space and the ensemble of its inter-subunit contacts and interactions, without regard to the internal geometry of the subunits'.** The interactions between the subunits of the aggregates are similar to those found in the tertiary structure. The centres of interface between these subunits primarily involve hydrophobic interactions between nonpolar side chains as contained in the interiors of the individual subunits while on the periphery the interactions involve hydrogen bonding and salt bridges between ionised side chains. Further, the subunits in all quaternary structures are highly complementary in terms of shape and interactions between polar groups.

The activities of some proteins require the presence of a prosthetic group, a small non-peptide molecule or metal that binds tightly to a protein, keeping the protein in a fixed conformation and participation in binding ligands.

In haemoglobin (Fig. 3.11) each of the four chains binds a prosthetic group called haeme, which consist of an iron atom held in a cage by protoporphyrin. The haeme groups are the oxygen binding components of haemoglobin.

Fig. 3.11 *Structure of haemoglobin showing four subunits and the haeme moiety.*

Many proteins are dimers, trimers, tetramers or higher order aggregates of identical subunits. For example, bacteriochlorophyll protein contains a trimer, tobacco mosaic virus contains a seventeenmer while an isosahedron virus has the largest symmetrical structure possible with as many as sixty identical subunits.

EXERCISES

1. Proteins show a great deal of functional diversity. Justify the statement.

2. How are proteins classified on the basis of their shape and structure?

3. Proteins perform such diverse range of functions that these can be classified on the basis of their biological functions. Briefly describe the classification of proteins on the basis of their biological functions.

4. What do you understand by the amphoteric nature of proteins? How this property of proteins is exploited in separating a mixture of proteins?

5. Clearly differentiate between the terms 'salting in' and 'salting out' in the context of protein coagulation.

6. What do you understand by denaturation of proteins? List some commonly used agents causing denaturation of proteins.

7. Ninhydin and fluorescamine both can be used for the detection of proteins and amino acids. In what way is fluorescamine better than ninhydrin?

8. In determining the composition of proteins by acid hydrolysis it is not possible to distinguish between glutamine and glutamic acid. Explain.

9. Determination of the sequence of amino acids in a given protein is crucial to the understanding of the structure of the proteins. List the steps involved in such a determination.

10. How can you establish the presence of inter and intramolecular disulphide linkages in a given protein? Explain.

11. List different methods for the determination of the N-terminal amino acid in a peptide or protein. In what way is Edman degradation method better than the other methods?

12. List different methods for the determination of the C-terminal amino acid in a peptide or protein. The presence of aspagine and glutamine at the C-terminal of the peptide cannot be ascertained by hydrazinolysis method.Comment.

13. Draw the structure of the products formed on reaction of the tripeptide; Val-Lys-Gly with the following reagents followed by hydrolysis.

    (a) Fluoro-2, 4-dinitrobenzene
    (b) Phenylisothiocyanate
    (c) Cyanogen bromide

14. Cyanogen bromide cleaves a peptide specifically at the carboxyl side of the methionine residue. Explain giving the mechanism for the cleavage.

15. Suggest a mechanism for the reaction of Sanger's reagent (fluoro-2, 4-dinitrobenzene) with a peptide.

16. What products would you get by cleaving the peptide hormone glucagon having the following sequence?

    H-His-Ser-Gln-Gly-Thr-Phe-Thr-Ser-Asp-Tyr-Ser-Lys-Tyr-Leu-Asp-Ser-Arg-Ala-Gln-Asp-Phe-Val-Gln-Trp-Leu-Met-Asn-Thr-OH

    by

    (a) Trypsin
    (b) Chymotrypsin
    (c) Cyanogen bromide

17. A mutant form of the peptide hormone angiotensin-II is found to be an octapeptide containing a residue each of Asp, Arg, Ile, Phe, Pro, Met, Tyr and Val. The following results were obtained on enzymatic hydrolysis.

Enzyme	Hydrolysis products
Trypsin	A dipeptide containing Arg and Asp and a hexapeptide
Carboxypeptidase	Phe and a heptapeptide
Chymotrypsin	Tetrapeptide A (containing Asp, Arg, Tyr and Val)
	Tetrapeptide B (containing Ile, Phe, Pro and Met)
Cyanogen bromide	A dipeptide containing Phe and Pro and a hexapeptide

Determine the sequence of the peptide.

Chapter 4

Enzymes

4.1 INTRODUCTION

Enzymes are proteins that act as biological catalysts, i.e., these alter the rates of biochemical reactions without undergoing any permanent change in themselves. These have a high degree of specificity besides high efficiency or rate of reactions. The term enzyme was introduced by W. Kühne in 1878 to refer to certain substances in yeast (in *zymin*) that made it act as a fermentative –an agent causing fermentation. In nature, enzymes help millions of chemical reactions to occur at extraordinary speeds and under moderate conditions. In the absence of enzymes most chemical reactions that maintain a living organism would occur only under very drastic conditions say at a temperature of the order of 100 degree or above which would 'kill' the fragile cell. At normal body temperatures, these reactions would often proceed at an extremely slow rate. For example, the dissociation of carbonic acid into CO_2 and H_2O that takes place in the lungs, proceeds only at a rate of about 10^{-7} mol dm^{-3} s^{-1} at room temperature in a test tube. On the other hand, in the cells, the enzyme *carbonic anhydrase* accelerates the reaction by more than a million times.

$$H_2CO_3 \xrightarrow{\text{\textit{carbonic anhydrase}}} CO_2 + H_2O$$

Similarly, a single molecule of the enzyme, *catalase*, catalyses the decomposition of about 10^7 molecules of hydrogen peroxide –a toxic by-product of metabolism, in one second. A comparison of the rates of the catalysed and uncatalysed reactions give an idea of the capability of the enzymes. For example,

$$2\,H_2O_2 \xrightarrow{\text{\textit{catalase}}} 2\,H_2O + O_2 \quad K_{295} = 3.5 \times 10^7$$

$$2\,H_2O_2 \xrightarrow{Fe^{2+}} 2\,H_2O + O_2 \quad K_{295} = 56$$

The macromolecular components of almost all enzymes are composed of proteins, except for a class of RNA modifying catalysts known as **ribozymes**. These are molecules of ribonucleic acid that catalyse reactions on the phosphodiester bond of other RNAs. Besides sustaining life, enzymes are important in industry also. These have been in use for ages in the production of wine, curd and cheese etc. In recent times the synthetic applications of enzymes have revolutionised the area of synthetic Organic Chemistry.

4.2 NOMENCLATURE AND CLASSIFICATION OF ENZYMES

Conventionally, the trivial or common names of the enzymes were derived from the names of the substrate they acted on or the reaction they catalysed or both and ended in a suffix '–ase'. For example, *urease* is an enzyme acting on urea, *dehydrogenase* catalyses the reaction involving removal of hydrogen and *lactate dehydrogenase* refers to an enzyme that catalyses the removal of hydrogen from the lactate ions to give pyruvate.

$$H_2N-\underset{\underset{Urea}{}}{\overset{\overset{O}{\|}}{C}}-NH_2 \ + \ H_2O \ \xrightarrow{\ urease\ } \ \underset{\underset{dioxide}{Carbon}}{CO_2} \ + \ \underset{Ammonia}{2NH_3}$$

$$\underset{Ethanol}{CH_3CH_2OH} \ + \ NAD^+ \ \xrightarrow{\ dehydrogenase\ } \ \underset{Ethanal}{CH_3CHO} \ + \ NADH$$

$$\underset{\underset{Lactate}{}}{HO-\underset{\underset{CH_3}{|}}{\overset{\overset{COO^-}{|}}{C}}-H} \ + \ NAD^+ \ \underset{\ }{\overset{lactate\ dehydrogenase}{\rightleftharpoons}} \ \underset{\underset{Pyruvate}{}}{\overset{\overset{COO^-}{|}}{\underset{\underset{CH_3}{|}}{C}}=O} \ + \ NADH \ + \ H^+$$

The system worked well initially with fewer enzymes but as the number of enzymes known increased, their naming in this way proved far from satisfactory. In many cases, the same enzyme was known by several different names, while sometimes same name was given to different enzymes. Further, many of the names conveyed little or nothing about the nature of the substrate or reactions catalysed, e.g., catalase, chymotrypsin etc. At times, similar names were given to enzymes of quite different types. To meet this situation, an International Commission on enzymes was set up by the **International Union of Biochemistry** (IUB) in 1957. The **Commission on Enzymes (EC)** considered the question of a systematic and logical nomenclature for enzymes and recommended two types of nomenclature; one systematic and the other working or trivial. The nomenclature has been revised and updated from time to time in consultation with IUPAC, earlier by the EC and later by the **Nomenclature Committee of International Union of Biochemistry and Molecular Biology (NC-IUBMB)**. The currently used nomenclature (NC-IUBMB; 1992 recommendations) is briefly described below.

4.2.1 Systematic and Recommended Names

The systematic name of an enzyme is given in accordance with definite rules and describes the action of an enzyme as exactly as possible thereby identifying the enzyme precisely. The **systematic name** of an enzyme ends in -*ase* and consists of two parts. The first part contains the name of the substrate, while the second part indicates the nature of the reaction e.g., *alcohol dehydrogenase*. However, in case of a bimolecular reaction the first part contains the names of both the substrates separated by a colon and has small and equal spaces before and after the colon. For example, for the bimolecular reaction,

ATP + Creatine ——————➤ ADP + Phosphocreatine

the systematic name of the enzyme is ATP: creatine phosphotransferase. The enzyme catalyses the transfer of a phosphate group from ATP to creatine. The trivial name for this enzyme is *creatine kinase*. Though systematic names are quite logical, these are quite cumbersome and inconvenient to use, therefore, the commission later decided to give more prominence to the trivial names. However, the systematic names are still retained as the basis for classification of enzymes (discussed below) which gives a code number to every enzyme. All the known enzymes are given code numbers depending on the their classes and sub-classes etc. These are collected in a list called as **enzyme list**, in which the common names follow immediately after the code number, and are described as **recommended names**. In assigning recommended names, a name in common use is taken as such if it gives some indication of the reaction and is not incorrect or ambiguous. Otherwise a recommended name is based on the same general principles as the systematic name but with only essential details so as to produce a name short enough for convenient use.

4.2.2 Classification Numbers and Code Names

The enzyme commission has devised a system for classification of enzymes that also serves as a basis for assigning code numbers to them. In this system each enzyme is classified by assigning an EC (enzyme commission) number to it which consists of four parts. Three general principles have been used by the commission for this purpose.

1. An enzyme is to be named by adding the suffix '**–ase**' to the name of the substrate on which it acts and refers only to single enzymes, *i.e.* single catalytic entities. In cases where a number of enzymes are involved in catalysing an overall reaction, the word *system* is to be included in the name. For example, the oxidation of succinate by molecular oxygen involves succinate dehydrogenase, cytochrome

oxidase, and several other intermediate carriers. These enzymes are named as a group called *succinate oxidase system* and not as *succinate oxidase*.

2. Since the chemical reaction catalysed is the specific property that distinguishes one enzyme from another, it is logical to use it as the basis for the classification and naming of enzymes. Therefore, according to the second general principle, the enzymes are principally classified and named according to the reaction they catalyse. For example, *dehydrogenases* catalyse a reaction involving removal of hydrogen.

3. According to the third guiding principle, the enzymes divided into groups on the basis of the type of reaction catalysed, along with the name(s) of the substrate(s) provides a basis for naming individual enzymes. For example, *lactate dehydrogenase* acts on the substrate lactate in catalysing the removal of hydrogen atoms.

An advantage of this method of classification is that the chemical reactions catalysed by the enzymes fall into relatively smaller number of types whereas, the large number of the available enzymes are due to their high specificity towards the part of the substrate molecule other than the group actually undergoing the reaction.

In the light of above mentioned principles, the enzymes have been classified into six main classes, which are further divided into sub-classes and sub-sub-classes. Presently more than 3500 different enzymes have been classified and assigned code numbers by the Nomenclature Committee of International Union of Biochemistry and Molecular Biology (NC–IUBMB). These code numbers, which are now widely in use, are prefixed by EC and contain four digits or figures (a.b.c.d) separated by points, for example, EC 1.3.2.6, with the following meaning:

(i) the first digit shows to which of the six main classes does the enzyme belong

(ii) the second digit indicates the sub-class

(iii) the third digit gives the sub-sub-class

(iv) the fourth digit is the serial number of the enzyme in its sub-sub-class

The six main classes and their division into sub-classes and sub-sub classes are described below.

Class 1

Oxidoreductases: The enzymes belonging to this class catalyse oxidation-reduction reactions. The substrate that is oxidised is regarded as hydrogen donor and the systematic name is created as donor : acceptor oxidoreductase. The recommended name will be a *dehydrogenase* or *reductase* (if transfer of hydrogen from the donor is not obvious) or *oxidase* (if O_2 is the acceptor) as the case may be. This class is assigned a number 1 or a code number EC 1, the second figure in the code number of the oxidoreductases indicates the group in the hydrogen donor which undergoes oxidation and the third figure indicates the type of acceptor involved. The last digit refers to the substrate in question.

Table 4.1 shows the sub-classes, sub-sub-classes for a sub-class (1.1) and the enzyme list (truncated) of the sub-sub-class (EC 1.1.3) for oxidoreductases (class 1). Let us take an example to understand this classification scheme. Suppose the code for an enzyme is EC 1.1.3.4. Here EC stands for Enzyme Commission, 1 stands for the main class i.e. *oxidoreductases*, next 1 stands for acting on CH–OH group of donors, 3 stands for oxygen as acceptor and 4 stands for the substrate in question i.e. β-D-glucose.

In fact this enzyme catalyses the following reaction.

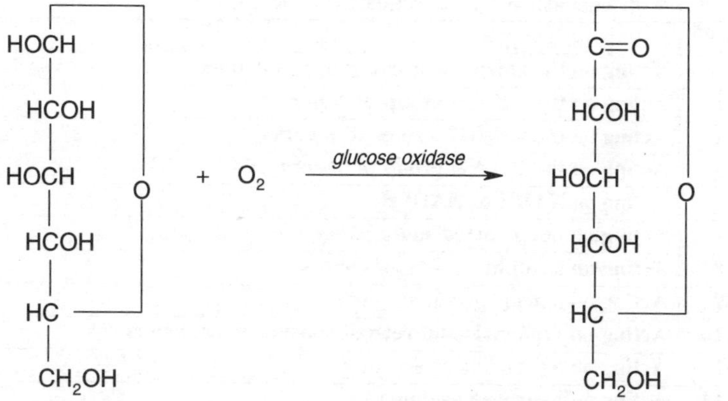

β-D-Glucose   + O_2 →(glucose oxidase)   δ-Glucono-1, 5-lactone

The systematic name given to this enzyme would be:

β-D-glucose : oxygen 1-oxidoreductase

As can be seen in the table, the recommended name for the enzyme is *glucose oxidase*.

Class 2

Transferases: Transferases are enzymes transferring a group, *e.g.* a methyl group, a glycosyl group or a phosphate group, from one compound (donor) to another compound (acceptor).

Table 4.1 The sub-classes, sub-sub-classes and the enzyme list (truncated) for the class, *oxidoreductases* (EC 1)

Number		Individual enzymes	Recommended names
EC 1.1	**Acting on the CH-OH group of donors**		
EC 1.1.1	With NAD or NADP as acceptor		
EC 1.1.2	With a cytochrome as acceptor		
EC 1.1.3	With oxygen as acceptor	EC 1.1.3.1	deleted, included in EC 1.1.3.15
		EC 1.1.3.2	now EC 1.13.12.4
		EC 1.1.3.3	malate oxidase
		EC 1.1.3.4	**glucose oxidase**
		EC 1.1.3.5	hexose oxidase
		EC 1.1.3.6	cholesterol oxidase
		EC 1.1.3.7	aryl-alcohol oxidase
		EC 1.1.3.8	L-gluconolactone oxidase
		EC 1.1.3.9	galactose oxidase
		EC 1.1.3.10	pyranose oxidase
EC 1.1.4	With a disulfide as acceptor		
EC 1.1.5	With a quinone or similar compound as acceptor		
EC1.1.99*	With other acceptors		
EC 1.2	**Acting on the aldehyde or oxo group of donors**		
EC 1.3	**Acting on the CH-CH group of donors**		
EC 1.4	**Acting on the CH-NH$_2$ group of donors**		
EC 1.5	**Acting on the CH-NH group of donors**		
EC 1.6	**Acting on NADH or NADPH**		
EC 1.7	**Acting on other nitrogenous compounds as donors**		
EC 1.8	**Acting on a sulphur group of donors**		
EC 1.9	**Acting on a heme group of donors**		
EC 1.10	**Acting on diphenols and related substances as donors**		
EC 1.11	**Acting on a peroxide as acceptor**		
EC 1.12	**Acting on hydrogen as donor**		
EC 1.13	**Acting on single donors with incorporation of molecular oxygen (oxygenases)**		
EC 1.14	**Acting on paired donors, with incorporation or reduction of molecular oxygen**		
EC 1.15	**Acting on superoxide as acceptor**		
EC 1.16	**Oxidising metal ions**		
EC 1.17	**Acting on CH$_2$ groups**		
EC 1.18	**Acting on iron-sulphur proteins as donors**		
EC 1.19	**Acting on reduced flavodoxin as donor**		
EC 1.20	**Acting on phosphorus or arsenic in donors**		
EC 1.97	**Other oxidoreductases**		

* In a few cases of sub-classes or sub-sub-classes the subdivision is given a number 99 and is designated as 'others' so as to leave space for the new subdivisions.

For example, the following reaction,

Adenosinetriphosphate (ATP)   D-Glucose

glucokinase

Adenosinediphosphate (ADP)   D-Glucose-6-phosphate

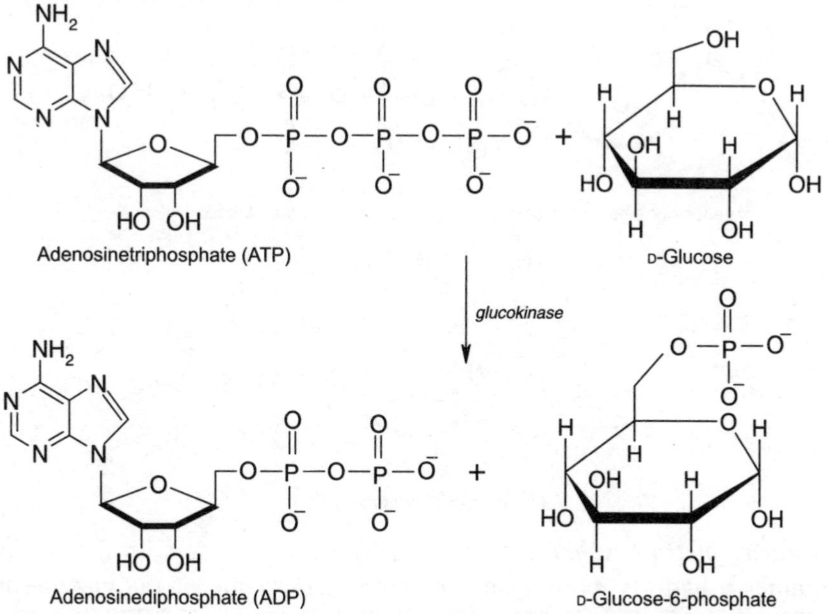

is an example of a reaction involving the transfer of a phosphate group. Here, ATP is the donor of phosphate group which is accepted by D-glucose. The **systematic names** of the enzymes belonging to this class are formed according to the scheme

donor : acceptor group-transferred-transferase

Thus, the full systematic name of the enzyme catalysing the above reaction would be:

ATP : D-glucose 6-phosphotransferase

This enzyme is commonly known as *glucokinase*. The EC code for this enzyme is 2.7.1.2. The first digit refers to the class transferases while the second figure in the code number indicates the group transferred (phosphate) and the third figure gives further information on the transferred group. The last digit refers to the substrate in question. The recommended names for this class are normally formed according to *acceptor grouptransferase* or *donor grouptransferase*.

Class 3

Hydrolases: These enzymes catalyse the hydrolytic cleavage of C–O, C–N, C–C and some other bonds including phosphoric anhydride bonds. Although the systematic name always includes *hydrolase*, the recommended name is, in many cases, formed by the name of the substrate with the suffix –*ase*. For example, the enzyme catalysing the following reaction,

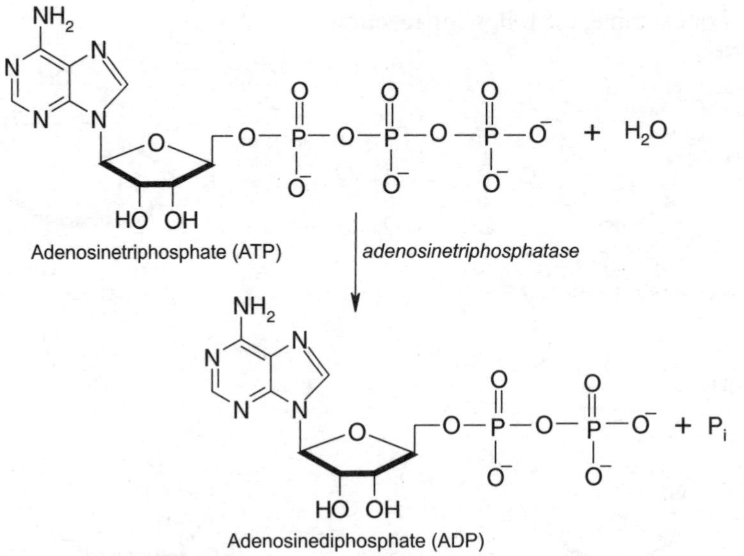

Adenosinetriphosphate (ATP)   *adenosinetriphosphatase*

Adenosinediphosphate (ADP)

is *adenosinetriphosphatase*. The name of the substrate with this suffix implies a hydrolytic enzyme. The systematic name of the enzyme is ATP : phosphohydrolase. The EC code for this enzyme is EC 3.6.1.3. The first digit as usual refers to the class, the second figure in the code number of the hydrolases indicates the nature of the bond hydrolysed and the third figure normally specifies the nature of the substrate. The last figure refers to the substrate (ATP in the given example) in question.

Class 4

Lyases: Lyases are enzymes cleaving C–C, C–O, C–N and other bonds by elimination, leaving double bonds or rings, or conversely adding groups to double bonds. The systematic name is formed according to the pattern substrate group-lyase. For example, the enzyme fructose-bisphosphate aldolase catalyses the following reaction.

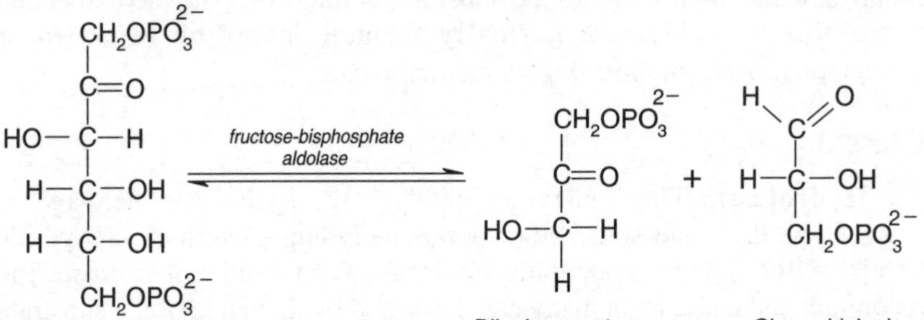

Fructose-1, 6-bisphosphate   Dihydroxyacetone phosphate   Glyceraldehyde-3-phosphate

The systematic name of the enzyme is D-fructose-1,6-bisphosphate D-glyceraldehyde-3-phosphate-lyase. The hyphen is an important part of the name, and to avoid confusion this hyphen should not be omitted, *e.g.* *phosphate-lyase* not '*phosphatelyase*'. The EC code for the enzyme is EC 4.1.2.13. In this the second figure indicates the bond broken and the third figure gives further information on the group eliminated. The last digit refers to the substrate in question.

Class 5

Isomerases: These enzymes catalyse geometric or structural changes within one molecule. According to the type of isomerism, they may be called *racemases, epimerases* or *tautomerases etc.* For example, triosephosphate isomerase catalyses the following reaction.

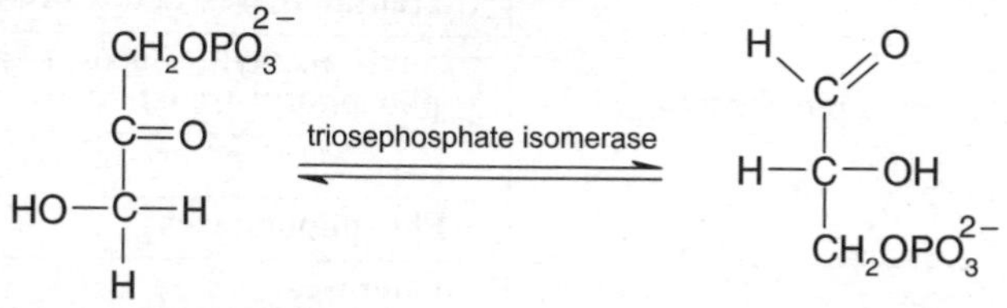

Dihydroxyacetone phosphate   Glyceraldehyde-3-phosphate

The systematic name of the enzyme is D-glyceraldehyde-3-phosphate ketol-isomerase and the code is EC 5.3.1.1. Here, the second digit refers to the type of isomerism while the third digit indicates the type of substrate. The last digit refers to the substrate in question.

Class 6

Ligases: Ligases are enzymes catalysing the joining together of two molecules coupled with the hydrolysis of a diphosphate bond in ATP or a similar triphosphate. The systematic names are formed on the system $X : Y$ *ligase* (*AMP or ADP-forming*) where X and Y are the two molecules to be joined together. For example, isoleucyl-t-RNA ligase catalyses the linking of L-isoleucine to t-RNA^{Ile} i.e., the isoleucine acceptor t-RNA.

L-Isoleucine + t-RNA^{Ile} + ATP $\xrightarrow{\text{isoleucine-t-RNA ligase}}$

L-Isoleucyl- t-RNA^{Ile} + AMP + pyrophosphate

The systematic name of the enzyme would be L-Isoleucine : t-RNA^{Ile} ligase (AMP-forming) and the code is EC 6.1.1.5. The second figure refers to the type of bond formed (carbon-oxygen in the present case) while third figure indicates the type of the compound formed. The last figure refers to the substrate.

Table 4.2 gives a summary of major sub-classes of different enzyme classes.

Table 4.2 Enzyme classes and their major sub-classes

Class No.	Enzyme class name	Major sub-classes
1.	Oxidoreductases	Dehydrogenases
		Oxidases
		Reductases
		Peroxidases
		Catalases
		Oxygenases
		Hydroxylases
2.	Transferases	Transaldolases and transketolases
		Acyl-, methyl-, glucosyl– and phosphoryl-transferases
		Kinases
		Phosphomutases
3.	Hydrolases	Esterases
		Glycosidases
		Peptidases
		Phosphatases
		Thiolases
		Phospholipases
		Amidases
		Deaminases
		Ribonucleases
4.	Lyases	Decarboxylases Aldolases
		Hydratases
		Dehydratases
		Synthases
		Lyases
5.	Isomerases	Racemases
		Epimerases
		Isomerases
		Mutases (not all)
6.	Ligases	Synthetases
		Carboxylases

4.3   CHARACTERISTICS OF ENZYMES

Most enzymes, as mentioned earlier, are proteins that are synthesised by living cells and act as catalysts for the large number of biochemical reactions in a cell. Enzymes display certain characteristic properties which are in contrast to those of their chemical counterparts that is, chemical catalysts. The significant characteristics are

- Extent of rate enhancement or catalytic power
- Specificity and
- Regulation

4.3.1 Catalytic Power

As mentioned in section 4.1, the enzyme catalysed reactions may have extremely high rates. Typically, enzymes **accelerate** the reactions by a factor of 10^6 to 10^{12} and allow millions of reactions necessary in our body to proceed in very short time. An example is provided by the enzyme *urease* that catalyses the hydrolysis of urea. Here the catalysed reaction has a rate that is roughly 13 orders of magnitude higher.

$$H_2N-\underset{\underset{\text{Urea}}{}}{\overset{\overset{O}{\|}}{C}}-NH_2 + H_2O \xrightarrow{\text{\textit{urease}}} CO_2 + 2NH_3; \quad K \sim 5 \times 10^6$$

$$H_2N-\overset{\overset{O}{\|}}{C}-NH_2 + H_2O \xrightarrow{H^+} CO_2 + 2NH_3; \quad K \sim 5 \times 10^{-7}$$

In extreme cases the increase in the rate could be as high as 10^{17} times. The enzymes achieve these high rates by altering the thermodynamics of a reaction in such a way that the reactants and products of a reaction reach equilibrium much faster than otherwise would, without altering their equilibrium concentrations. The enzymes, like chemical catalysts, are not altered during their reactions and are released unchanged after catalysing the reaction. Therefore, these are required only in small amounts in the cell.

4.3.2 Enzyme Specificity

Enzymes show a high degree of specificity towards their substrates, the products and to the type of chemical reaction being catalysed as compared to their chemical counterparts. For example, the rate of oxidation of β-D-glucose by *glucose oxidase* is about 150 times faster than that for the α-anomer. Further, if a given substrate can give two different types of products then it would be acted over by two different enzymes, each giving a specific product. Formation of a side product is very rare in an enzyme catalysed reaction. Thus, *specificity of enzymes refers to the ability of an enzyme to selectively recognise and transform a given substrate out of a mixture of a number of substrate types.*

Different enzymes demonstrate different types and extents of specificities. These can arbitrarily be put into following groups.

- Absolute specificity
- Stereochemical specificity
- Linkage specificity and
- Group specificity

Absolute Specificity

As the name suggests certain enzymes specifically catalyse a given reaction on a particular substrate. *Urease* is an example of this class as it catalyses the hydrolysis of urea only and does not show any reaction with closely related substrates like, thiourea or methylurea.

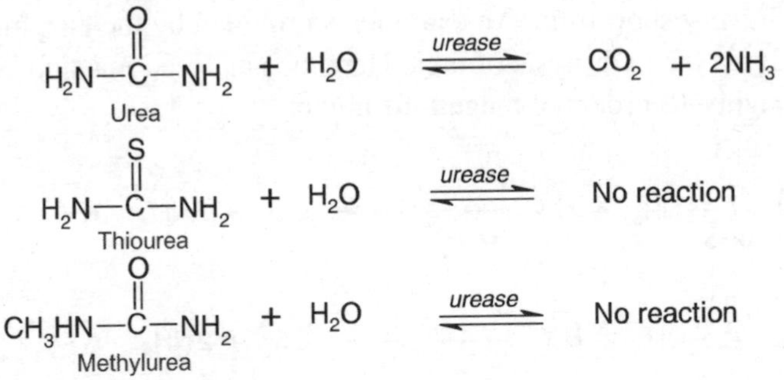

$$H_2N-\underset{\substack{\| \\ O}}{C}-NH_2 \ + \ H_2O \ \underset{}{\overset{urease}{\rightleftharpoons}} \ CO_2 \ + \ 2NH_3$$
Urea

$$H_2N-\underset{\substack{\| \\ S}}{C}-NH_2 \ + \ H_2O \ \underset{}{\overset{urease}{\rightleftharpoons}} \ \text{No reaction}$$
Thiourea

$$CH_3HN-\underset{\substack{\| \\ O}}{C}-NH_2 \ + \ H_2O \ \underset{}{\overset{urease}{\rightleftharpoons}} \ \text{No reaction}$$
Methylurea

Stereochemical Specificity

Many enzymes show specificity towards the stereochemical nature of the substrate. For example, enzyme *fumarase* catalyses addition of water to fumarate (*trans* isomer) to give malate where as maleate (the *cis* isomer) remains unaffected.

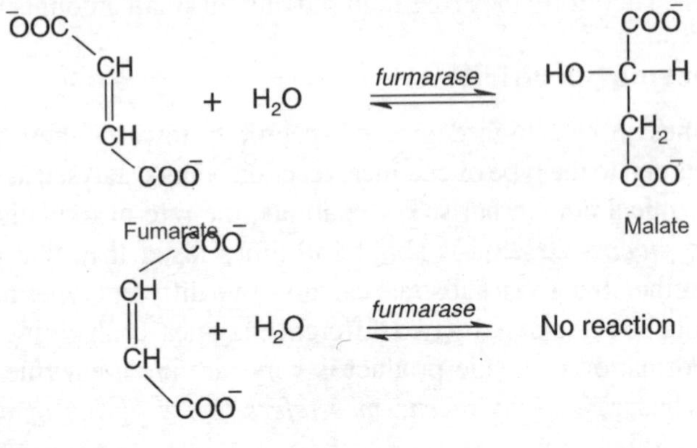

Linkage Specificity

Some enzymes have specificity towards a kind of bond in the substrate rather than the substrate as a whole. For example, *peptidases* are specific to the peptide bond.

Group Specificity

Some enzymes however show group specificities i.e. acting on a group instead of a particular bond or substrate. For example, the enzyme *carboxypeptidase A* selectively removes the C-terminal amino acid in a peptide containing a free C-terminal or *pepsin* hydrolyses peptide bonds having adjacent aromatic amino acids.

As discussed earlier, the specificity of enzymes has provided a convenient system for classifying and naming them. The enzymes are sensitive to the changes in pH, temperature and nature and concentration of salt and work best under optimum conditions (Sec. 4.5).

Enzymes have a somewhat broad range of substrate specificity, i.e., a given substrate may be acted upon by a number of different enzymes, each of which uses the same substrate(s) and produces the same product(s). The individual members of a set of enzymes acting at same substrate and producing same product are known as **isozymes**. Isozymes are a set of closely related enzymes that are produced genetically in the organism. These differ slightly in terms of their primary structure (the amino acid sequence) or higher order (conformation) structures. Sometimes these may differ even in terms of certain covalent modifications. In a set of isozymes the individual members are distinguished on the basis of their electrophoretic mobilities. These are designated by numbers, the most mobile isozyme is given number 1 and the others follow in the sequence. Various isozymes of a group are often found in different tissues of the body. The best studied set of isozymes is the *lactate dehydrogenase* **(LDH) system.** A number of genetically related forms of *lactate dehydrogenase* are found in the serum and muscle extracts.

4. 3.3 Enzyme Regulation

In addition to the specificity, the capacity to be **regulated** is another important feature of enzymes. The catalytic activity of many enzymes is found to depend on the concentrations of the substances other than their substrates. These substances may be small molecules or ions or substrate analogues or at times totally unrelated to the substrate. This characteristic of the enzymes is extensively exploited by the cell to make the most effective use of the cellular enzymes.

4.4 MECHANISM OF ENZYME ACTION

In general, catalysts increase the rate of product formation by lowering the activation energy for the reaction and facilitating favourable orientation of colliding reactant molecules for product formation. The mechanisms by which enzymes lower the energy of activation are still not totally understood. However, the mechanisms are believed to be directly or indirectly related to achievement of a **transition state** for the reaction. A portion of the overall tertiary structure of the enzyme is responsible for its activity and is called **active site** (Fig 4.1). The active site is quite small as compared to the overall size of the enzyme. The side chains of some of the amino acids in the active site help in holding the substrate to the enzyme and are called **binding groups,** while side chains of other amino acids are involved in the catalytic process and are called **catalytic groups**. For example, in the enzyme *trypsin*, its complex tertiary structure brings together a histidine residue from one section of the molecule with glycine and serine residues from another. The side chains of the residues in this particular geometry produce the site that accounts for the enzyme's reactivity. The catalytic activity of most of the enzymes can be accounted for by the presence of five functional groups given in Table 4.3. It is not necessary that the active sites of all the enzymes contain all the five functional groups.

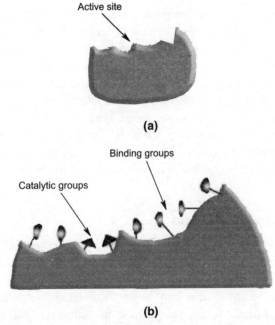

Active site

(a)

Binding groups

Catalytic groups

(b)

Fig. 4.1 *(a) A schematic representation of the active site of an enzyme (b) A part of active site showing the binding and catalytic groups.*

Table 4.3 Catalytically active functional groups

Functional group	Amino acid	pK_a
COOH, COO$^-$	Asp, Glu	4
OH	Ser	14–15
SH	Cys	10
NH$_3^+$	Lys	10
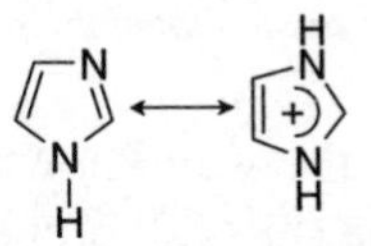	His	7

Binding of substrates to the active site of the enzyme brings them close together, thereby raising their effective concentration in the active site to many times the concentration in the surrounding solution. In this enzyme-substrate complex the polar and nonpolar groups of the active site may also bring the substrate molecules into an arrangement in which they can collide at the correct positions and orientations so as to form the transition state. This facilitates the bond breaking and making, leading to conversion of the reactants to the products. Once the reaction is complete, the product molecule separates itself from the active site and makes the site available for another incoming substrate molecule. This process occurs in a highly efficient manner measured by **turnover number** of the enzyme which refers to the number of molecules of substrate upon which a given molecule of the enzyme acts per second. These numbers can vary over about five orders of magnitude. The turnover numbers of some enzymes is compiled in Table 4.4.

Table 4.4 Turnover numbers of some enzymes

Enzyme	Turnover number (s^{-1})
DNA polymerase	1.5×10^1
Chymotrypsin	1.0×10^2
Lactate dehydrogenase	1.0×10^3
3-Ketoisomerase	2.8×10^5
Carbonic anhydrase	6.0×10^5

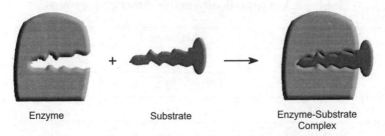

Enzyme   Substrate   Enzyme-Substrate
Complex

Fig. 4.2 *A diagrammatic representation of the lock and key model for enzyme specificity. The structure of the active site of the enzyme is complimentary to that of the substrate.*

The mechanism described above is based on the well known **"lock and key"** model for the specificity of enzyme action (Fig 4.2) proposed by Emil Fischer in 1890. The key (substrate) has a specific shape (arrangement of functional groups and other atoms) that allows it and no other key to fit into the lock (the active site of enzyme). Thus, the enzyme catalysed reactions proceed through a number of steps. These are illustrated in Fig. 4.3.

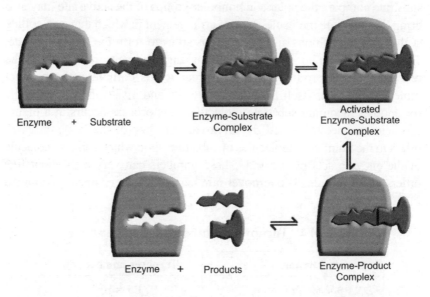

Enzyme   +   Substrate   Enzyme-Substrate
Complex   Activated
Enzyme-Substrate
Complex

Enzyme   +   Products   Enzyme-Product
Complex

Fig. 4.3 *A diagrammatic representation of various steps in enzyme catalysed reactions according to lock and key model.*

This model assumes the structure of the enzyme to be rigid. However, the developments in the understanding of protein structure suggest it to be dynamic in nature. To accommodate the flexibility of protein structure, the lock and key model was modified by Daniel E. Koshland Jr. in 1958. According to the **induced fit** or **hand and glove model** proposed by

Koshland the active site of the enzyme is not rigid but has a certain amount of flexibility whereby it can expand or contract to some extent so as to accommodate the substrate molecule. That is, when the substrate molecule approaches the enzyme, the active site acquires a shape complimentary to that of the substrate as shown in Fig 4.4. It is somewhat like a hand fitting into a glove. The glove adjusts in shape and size to fit different hands within a certain range of sizes. Further, in the E–S complex the enzyme forces the substrate molecule to be distorted to acquire a transition state. In other words, the induced fit model envisages changes in the enzyme as well as the substrate. It may be noticed here that the two models differ only in terms of the mechanism of E–S complex formation; the other steps in the catalytic mechanism remain the same.

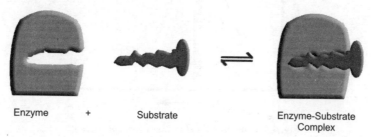

Enzyme   +   Substrate   Enzyme-Substrate
                                                                  Complex

Fig. 4.4 *A diagrammatic representation of the induced fit model for enzyme specificity. The active site of the enzyme gets 'induced' to take a shape complimentary to that of the substrate.*

4.5 FACTORS AFFECTING ENZYME ACTION

The activity of enzymes is quite sensitive to the factors like, temperature, pH, concentration of substrate, enzyme itself and the salt etc. These are discussed below.

Temperature

Every enzyme has an optimum range of temperature (usually ranging from about 30°C to 40°C) wherein it shows its maximum activity. The activity of the enzyme decreases or is completely lost outside this range. This occurs because the changes in temperature disrupt the forces stabilising the enzyme (protein) structure, which in turn may alter the active site to an extent that it is unable to accommodate the substrate molecules. At low body temperature (hypothermia) or very high body temperatures (hyperthermia), most of the enzymes within the human cells may lose activity or stop functioning and thereby hamper the functioning of the cell leading to severe physiological consequences.

Hydrogen Ion Concentration (pH)

The enzyme catalysed reactions are strongly influenced by the hydrogen ion concentration or the pH of the reaction mixture. Usually small changes in the pH of the cell can also affect the normal functioning of the enzyme. The pH dependence of an enzyme catalysed reaction is given in Fig.4.5. The bell shaped curve shows that the activity of the enzyme is maximum over a very narrow range of pH. This is called **optimal pH** range. The enzyme activity decreases at pH values lower or higher than this range.

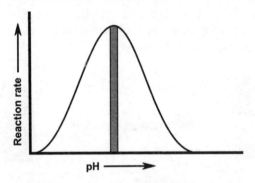

Fig. 4.5 *The effect of pH on the activity of enzyme.*

Most of the enzymes have an optimal pH range around 7.0; however, some enzymes can have their optimal pH in extremely acidic or basic range. For example, the enzymes like *pepsin* and chymotrypsin present in the stomach can operate effectively at a very low pH. While the enzymes like α-*amylase* found in the saliva of the mouth operate most effectively near neutral pH. On the other hand, certain enzymes like the *lipases* function most effectively at basic pH values. The optimal pH values for some of the common enzymes are given in Table 4.5.

Table 4.5 The optimal pH values of some common enzymes

Enzyme	Optimal pH value
α-*Amylase*	7.0
Alkaline phosphatase	9.5
Carboxypeptidase	7.5
α-*Glucosidase*	5.4
Pepsin	1.5
Trypsin	7.8
Urease	6.7

The change in pH from the optimal value can affect the structure of the enzymes by altering the ionic state of different side chains of the enzyme molecules. This in turn may alter the structure of the active site which as a consequence can no longer accommodate the substrate. Alternatively, the amino acid side chains in the active site itself may have its ionic state altered. In case of metabolic disorders leading to **acidosis** (a decrease in the blood pH) or **alkalosis** (an increase in the blood pH), the enzymes may lose their activity leading to undesired consequences. Correcting pH or temperature imbalances usually allows the enzyme to resume its original shape and function.

Concentration of the Enzyme

In presence of a sufficient concentration of the substrate, increasing enzyme concentration will increase the reaction rate, as more active sites are available to the substrate molecules.

Concentration of the Substrate

At a given concentration of the enzyme, the reaction rate is controlled by the substrate concentration at lower substrate concentrations. The rate of the reaction increases with an increase in the concentration of the substrate. However, at high concentrations of the substrate the enzyme gets saturated with the substrate and any increase in the concentration of substrate does not increase the reaction rate any further. A typical curve showing the variation of the reaction rate with an increase in the concentration of the substrate is given in Fig. 4.6.

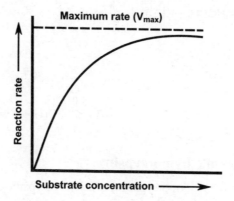

Fig. 4.6 *Variation of reaction rate with concentration of the substrate at a given concentration of enzyme.*

Salt Concentration

Each enzyme has an optimal working salt concentration. Changes in the salt concentration may also denature the enzyme and hamper its performance.

4.6 CHYMOTRYPSIN: AN ENZYME IN ACTION

Chymotrypsin is a proteolytic enzyme belonging to a broad group of enzymes called *serine proteases* that use **serine** side chain as a reactive nucleophile in the catalysed reaction. Chymotrypsin specifically cleaves the peptide bond next to an aromatic side chain and to a lesser extent the peptide bond next to a hydrophobic side chain like, methionine or leucine. Since *proteases* are destructive in nature these are normally synthesised in their inactive form called **zymogen**. These are suitably activated as per the requirements. Chymotrypsin is synthesised in the mammalian pancreas as an inactive precursor called **chymotrypsinogen**. This precursor is secreted in the intestine where it is activated by proteolytic cleavage by *proteases*. The structure and mechanism of action of chymotrypsin is probably the most extensively studied and understood system. It catalyses the hydrolysis of the amide (or peptide) bond with the help of a strong nucleophile in the form of $-CH_2OH$ group of a specific serine residue. The overall hydrolysis is split into two steps. In the first step an acyl-enzyme ester is formed as an intermediate, which is then hydrolysed by water in the second step to yield free carboxylic acid and the enzyme is regenerated. The detailed mechanism is explained later (Sec. 4.6.3).

$$R-\overset{\overset{\displaystyle O}{\|}}{C}-NHR' + HOCH_2-\boxed{Enz.} \xrightarrow{\text{Step 1}} R-\overset{\overset{\displaystyle O}{\|}}{C}-OCH_2-\boxed{Enz.} + R'NH_2$$

Peptide ———————————————— Acyl-enzyme ester

$$\Big\downarrow \text{Step 2 | Hydrolysis}$$

$$R-\overset{\overset{\displaystyle O}{\|}}{C}-OH + HOCH_2-\boxed{Enz.}$$

4.6.1 Structure of Chymotrypsin

The precursor to chymotrypsin (i.e., chymotrypsinogen) consists of a 245-residue long single chain polypeptide, which is held in its natural conformation by five intramolecular disulphide linkages. The linkages are between cysteine residues 1 and 122; 42 and 58; 136 and 201; 168 and 182 and 191 and 220. A schematic structure of chymotrypsinogen showing the disulphide linkages and cleavage sites is given in Fig 4.7.

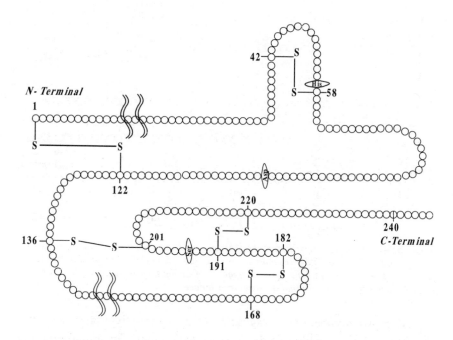

Fig. 4.7 *Schematic representation of the structure of chymotrypsinogen. Each circle represents an amino acid, three amino acids directly involved in the activity of chymotrypsin are shown as ovals. The vertical wavy lines indicate the positions where the chain is cleaved to get the active enzyme.*

Chymotrypsinogen is converted into the active enzyme in a sequence of interesting proteolytic cleavages. This process is initiated by an enzyme called *enterokinase* that specifically activates trypsinogen which in turn produces trypsin. Trypsin then executes the first cleavage (between residues Arg-15 and Ile-16) of chymotrypsinogen. This yields a two chain active enzyme called **π-chymotrypsin**. It digests itself in terms of a secondary cut between Leu-13 and Ser-14 residues to generate what is called **δ-chymotrypsin**. It is followed by two more cleavages (between Tyr-146 and Thr-147 and between Asp-148 and Ala-149) to give the active **α-chymotrypsin**. α-Chymotrypsin consists of three polypeptide chains held in place by three intramolecular and two intermolecular disulphide linkages as shown in Fig.4.8.

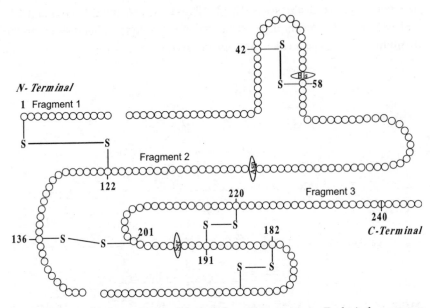

Fig. 4.8 *Schematic representation of the structure of chymotrypsin. Each circle represents an amino acid. The ovals indicate three amino acids directly involved in the activity of chymotrypsin. The three fragments of chymotrypsin have slightly different conformations than that in chymotrypsinogen.*

Though a number of cleavages take place in the activation process, it is the cleavage after residue 15 that is the most crucial. The amino group of the Ile-16 residue is very important as it gets into the formation of a salt link with the Asp-194 side chain. Such a possibility does not exist in the zymogen. The salt link in turn introduces conformational changes in the fragment 189-194 whereby the backbone amide bond of Gly-193 residue acquires an important position from the mechanism point of view.

4.6.2 Important Amino Acid Residues of Chymotrypsin

In terms of the activity of chymotrypsin three amino acids viz., Ser-195, His-57 and Asp-102 are the most important; serine-195 being the central residue. In addition, Ile-16 also has a role to play. Other residues provide the right geometry for its action.

Role of Serine-195

It is interesting to note that chymotrypsin contains as many as 28 serine residues but only one of these i.e., Ser-195 has a direct role in the activity of the enzyme. This has been established on the basis of the reaction of chymotrypsin with diisopropylfluorophosphate (a nerve gas) that leads to the loss of activity of the enzyme. This inhibitor reacts specifically and irreversibly by forming a covalent linkage with the side chain of Ser-195 residue only. The enzyme – inhibitor complex contains a phosphate ester having a structure similar to that of an acyl enzyme.

Diisopropylfluorophosphate   Diisopropyl-phosphate derivative of enzyme

This specificity is due to the spatial proximity of His-57 and Asp- 102 residues with serine-195 as revealed by X-ray crystallography. The specificity of Ser-195 was further established by reacting chymotrypsin with *p*-nitrophenylacetate followed by end group analysis.

Role of His-57

The importance of His-57 in the enzyme activity was suggested by the inhibition of its activity by a phenylalanine derivative containing reactive chloromethylketone group.

In the primary sequence of the enzyme, His-57 and Ser-195 are quite far from each other but in the tertiary structure these two residues are in close proximity. Affinity labelling studies also point out to the same. The orientation of His -57 side chain is such that the serine – OH group can form a hydrogen bond with the imidazole group of the histidine residue.

Role of Asp-102

Asp-102 residue in chymotrypsin is buried inside the globular fold of the protein –a rare energetically unfavourable feature in proteins. It is so oriented that it comes closer to the imidazole ring of His-57 residue from the side opposite to that of Ser-195. The histidine residue is thus flanked between aspartic acid and serine. The carboxylate ion in the aspartic acid side chain interacts with the nitrogen proton on the imidazole ring.

Site directed mutagenesis suggest that these three amino acid residues (Asp-102, His-57 and Ser-195) act synergistically to increase the rate of the enzyme catalysed reaction. The combination of His-57 and Asp-102 residues with Ser-195 create what is called a **charge-transfer relay system** or **catalytic triad.** These three residues are so arranged in space that these lead to a partial ionisation of the hydroxyl group in the Ser-195 side chain.

The interaction of carboxylate group of Asp-102 with imidazole side chain facilitates the ionisation of serinyl – OH group in the charge relay system as shown below.

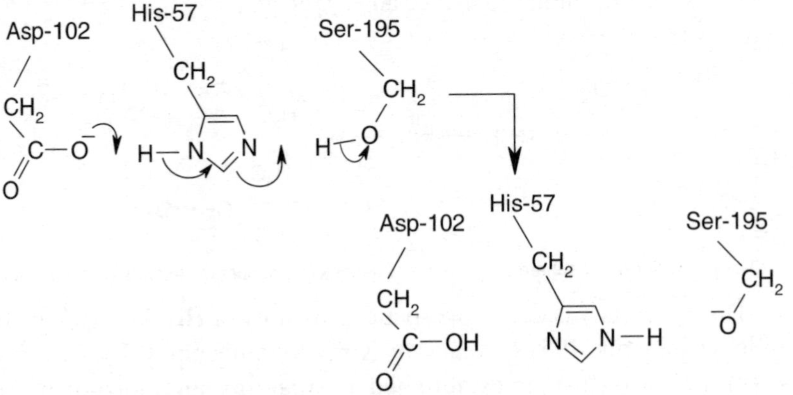

Fig. 4.9 *Charge-transfer relay system or catalytic triad operative in chymotrypsin and some other serine proteases like, trypsin etc.*

Substrate Binding Pocket

In addition to the three residues discussed above a few more residues are also important as these aid in the binding of the substrate to the enzyme. The binding site (or pocket) contains a number of nonpolar side chains, which favourably accommodates a hydrophobic side chain (e.g., Tyr, Trp, Phe etc.). The bottom of the binding pocket is occupied by Ser-189 while two glycyl residues (Gly-216 and Gly-226) are located near the opening of the pocket.

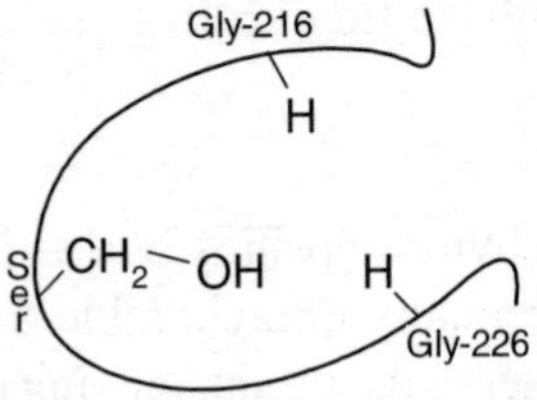

Fig. 4.10 *Substrate-binding pocket of chymotrypsin. A serine residue lies at the bottom of the pocket while two glycyl residues are located near the neck region.*

4.6.3 Mechanism of Action

The mechanism of action of chymotrypsin has been well studied. The sequence of steps showing the mechanism is depicted in Fig. 4.11 (a-g) along with their explanation.

When the substrate (a peptide or protein containing an aromatic or highly hydrophobic side chain) approaches the enzyme, the aromatic residue slips into the binding pocket of the enzyme where it is held in place by hydrophobic interactions with the residues constituting the pocket. In this enzyme-substrate complex, the carbonyl group of the peptide bond to be hydrolysed gets into hydrogen bonding with the peptide –NH groups of Ser-195 and Gly-193 residues. These two amide groups form what is referred to as **oxyanion binding site** or **hole**.

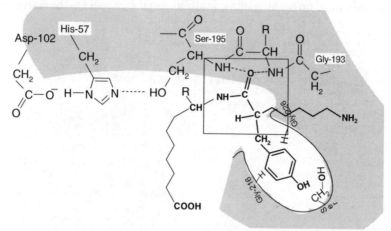

Fig. 4.11(a) *Substrate-chymotrypsin complex; the relevant side chains of the enzyme molecule are shown schematically. The schematic sequence shown in dark colour and smaller font represents the peptide substrate. The aromatic amino acid side chain of tyrosine is in the binding pocket of the enzyme.*

The reaction is then initiated by the nucleophilic attack of the Ser -195 side chain that carries a negative charge due to the charge relay system created by the participation of Asp -102 and His -57 side chains as explained earlier.

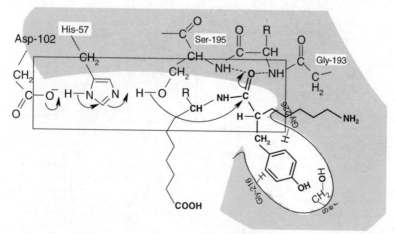

Fig. 4.11(b) *Nucleophilic attack of Ser -195 facilitated by the participation of Asp -102 and His -57 residues in the charge-transfer relay system.*

This leads to the formation of a tetrahedral intermediate, which is stabilised by the amide protons of Ser-195 and Gly-193. These amides stabilise the intermediate by accommodating the negative charge on the carbonyl carbon in the so-called 'oxanion'. The tetrahedral intermediate has not been observed directly, its existence is attributed by analogy to the tetrahedral adducts obtained on binding of the inhibitors to the enzyme in the active site.

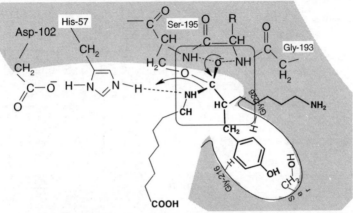

Fig. 4.11(c) *Formation and stabilisation of the tetrahedral intermediate. The negatively charged oxanion gets into hydrogen bonding with amide protons of Ser-195 and Gly-193.*

This short lived intermediate is decomposed in an acid catalysed elimination of a peptide fragment with a new N-terminal. In this process another intermediate is obtained in which the serine residue is linked to the enzyme as an acyl ester. Here, histidine residue acts as an acid and aids the formation of the acyl intermediate.

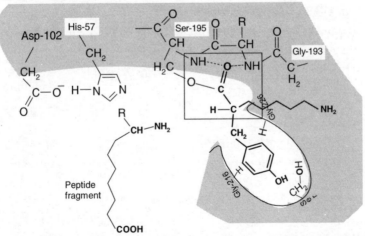

Fig. 4.11(d) *Formation of acyl-enzyme intermediate accompanied by elimination of a peptide fragment with a new N-terminal.*

Upto this stage, half of the reaction is over. The acyl-enzyme intermediate undergoes deacylation by a nucleophilic attack of a water molecule. The sequence of reactions is reverse of the ones explained above. The attack of water is facilitated by the participation of aspartate and histidine side chains.

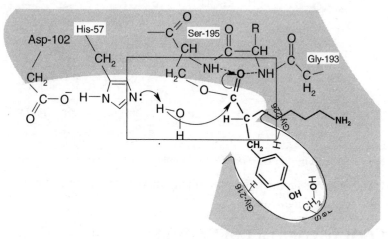

Fig. 4.11 (e) *Nucleophilic attack of water molecule facilitated by aspartate and histidine side chains.*

The deacylation also proceeds through the formation of a tetrahedral intermediate.

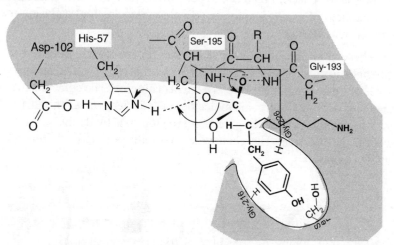

Fig. 4.11 (f) *Formation of second tetrahedral intermediate.*

This tetrahedral intermediate breaks down to yield an enzyme-product complex in which the second fragment of the protein (or peptide) having a new C-terminal is bound to the binding pocket of the enzyme.

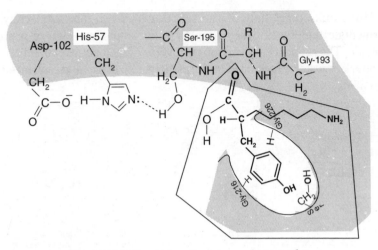

Fig. 4.11 (g) *Formation of enzyme-product complex.*

The enzyme-product complex finally dissociates to give free enzyme and the peptide fragment.

4.7 COFACTORS (or COENZYMES)

As discussed earlier, enzymes perform their functions with the help of functional groups of the amino acid side chains present in their active sites. However, these side chain functional groups are not capable of catalysing all the reactions needed by a cell to perform its functions. Certain enzymes need the help of small nonprotein organic molecules for catalysing these essential reactions.

These nonprotein groups, if covalently linked are referred to as **prosthetic groups**. The protein part in this combination is called **apoenzyme** while the other component is referred to as **cofactor**. The cofactors may be metal ions or complex organic molecules called **coenzymes**. The coenzyme binds to the apoenzyme to give **holoenzyme** that binds the substrate to perform the reaction. The coenzymes normally are derived from water soluble vitamins.

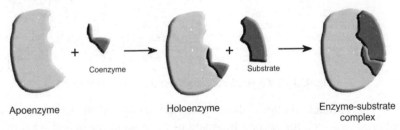

Fig. 4.12 *The role of coenzyme in a stable complex formation between the enzyme and substrate.*

Enzymes of IUB classes 1,2,5 and 6, catalysing oxidoreductions, group transfers, isomerisations and ligation reactions respectively often need coenzymes. The coenzymes may either be attached by covalent bonds to the enzyme (prosthetic group) or may exist freely in solution. Enzymes that require a metal in their composition are known as **metalloenzymes** if they bind and retain their metal atom(s) under all conditions, that is with very high affinity for the metal ion. Those enzymes which have a lower affinity for metal ion but still require the metal ion for their activity are known as **metal-activated enzymes**. Some of the metal ions commonly participating as cofactors for enzymes are given in Table 4.6.

Table 4.6 Important metal ion cofactors

Enzyme	Metal ion cofactor	Function of the metal ion
Cytochrome oxidase	Fe	Oxidation and reduction
Alcohol dehydrogenase	Zn	Binding of coenzyme; NAD$^+$
Urease	Ni	Part of catalytic site
Glutamate mutase	Co	Part of the coenzyme, cobalmin
Ascorbic acid oxidase	Cu	Oxidation and reduction

The coenzymes perform two important functions. Firstly, on binding to the enzyme these create an active site suitable for the substrate molecule to bind and secondly provide some functional groups to aid in the catalytic action of the enzyme. The functional role of coenzymes is to act as transporters of chemical groups from one reactant to another. The chemical groups carried can be as simple as the hydride ion ($H^+ + 2e^-$) carried by NAD$^+$ (nicotinamide adenine dinucleotide) or the molecule of hydrogen carried by FAD (flavin adenine dinucleotide); or they can be even more complex than the amine ($-NH_2$) carried by pyridoxal phosphate. These coenzymes are often altered structurally in the course of these reactions, the change being exactly opposite to the one taking place in the substrate. However, these are always restored to their original form in subsequent reactions catalysed by other enzyme systems.

Human body is unable to synthesise most of the coenzymes from elementary precursors available in the cell and need dietary support in terms of vitamins etc. All vitamins are not coenzymes per se, however, some are precursors that are transformed into coenzymes after metabolic modifications. Constitutionally, most coenzymes are obtained by modification of the water soluble vitamins ingested in food. If a given vitamin were not available in the diet, a deficiency with obvious clinical symptoms would arise.

Since coenzymes are chemically changed as a consequence of enzyme action, these are often considered to be a special class of substrates, or **second substrates**, which are common to many different holoenzymes. In all cases, the coenzymes donate the chemical group carried from the substrate to an acceptor molecule and get regenerated to their original form. This regeneration of coenzyme (unlike the usual substrates, which are consumed during the course of a reaction) and holoenzyme fulfills the definition of an enzyme as a chemical catalyst. The cofactors assist the enzyme in a number of ways. For example

- as inter-enzyme carriers
- as intermediates
- as templates or as primers in DNA synthesis
- as modifiers of the shape of enzymes
- as stabilisers of enzymes etc.

Some of the cofactors act in more than one way in a reaction. Of the various possible modes of action, the most common mode of cofactor action is as inter-enzymic carrier. In this mode, the carrier combines with the enzyme and and a part of the substrate is transferred to the carrier. This loaded carrier then migrates to another enzyme to transfer the load to the substrate on the other enzyme and then the cofactor dissociates from the enzyme. These inter-enzymic carriers can be classified in terms of the species or the group being carried. Some of the important coenzymes and their mode of action is described below.

4.7.1 Nicotinamide Adenine Dinucleotide (NAD⁺) and Nicotin- amide Adenine Dinucleotide Phosphate (NADP⁺)

NAD⁺ and NADP⁺ are two closely related coenzymes, derived from niacin (nicotinic acid and nicotinamide) or vitamin B_3. NAD⁺ is a dinucleotide containing nicotinamide, adenosine, ribose sugar and phosphoric acid while NADP⁺ contains an additional phosphate group in the 2′ position of ribose sugar attached to adenine.

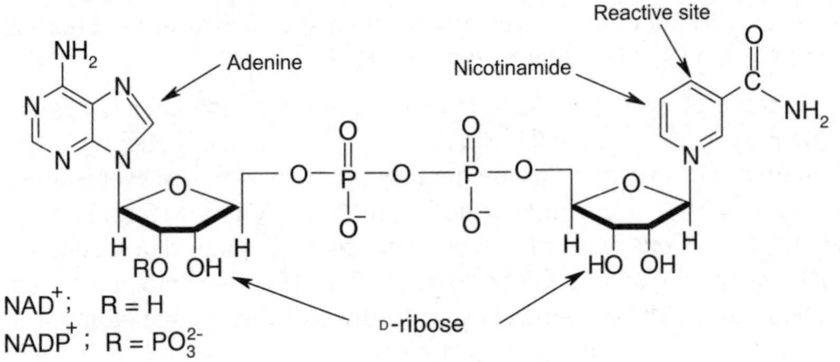

Structure of NAD⁺ / NADP⁺

Nicotinatamide is the amide of nicotinic acid which is closely related to nicotine, an alkaloid found (~2 to 8 % of dry weight) in tobacco leaves. Nicotinic acid can be obtained by the oxidation of nicotine.

Nicotine   Nicotinic acid

The deficiency of vitamin B_3 may result in physiological and psychological disorders. Mild deficiency leads to headache and dizziness whereas in case of acute deficiency there may be metabolic disorders like diarrhoea or even neurological symptoms like hallucinations.

These coenzymes perform the role of accepting equivalent of a hydride ion (a proton and two electrons) from the substrate to affect its oxidation. The nicotinamide ring is the functional part of the coenzymes; the para position (C-4) of the ring being the site of hydrogen transfer. As a consequence of hydrogen transfer the positive charge on the nitrogen atom is neutralised, the pyridine ring is reduced and loses its aromaticity. This gives a compound called 1,4-dihydronicotinamide adenine dinucleotide, NADH (or NADPH).

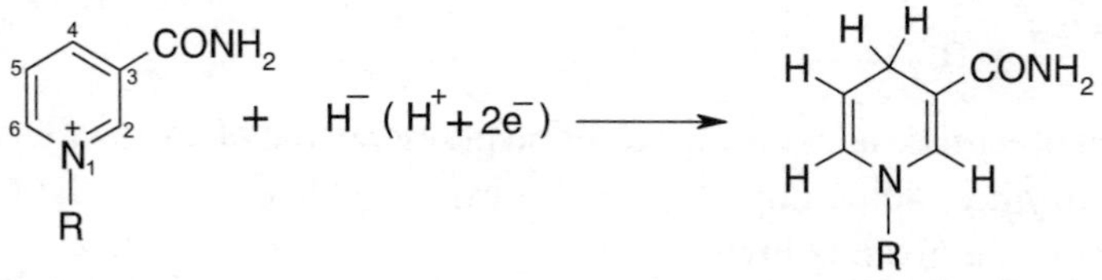

Nicotinamide   1,4-dihydronicotinamide
NAD$^+$ NADH

R = remaining part of the coenzyme molecule

The general reaction of NAD$^+$ with the substrate (AH$_2$) may be summarised as:

$$NAD^+ + AH_2 \rightleftharpoons NADH + H^+ + A$$

The transfer of hydrogen between the substrate and NAD$^+$ is direct and occurs with stereospecificity. The two hydrogens at C-4 position of NADH are chemically equivalent; it is the difference in their topology that is responsible for the stereospecificity of the reaction.

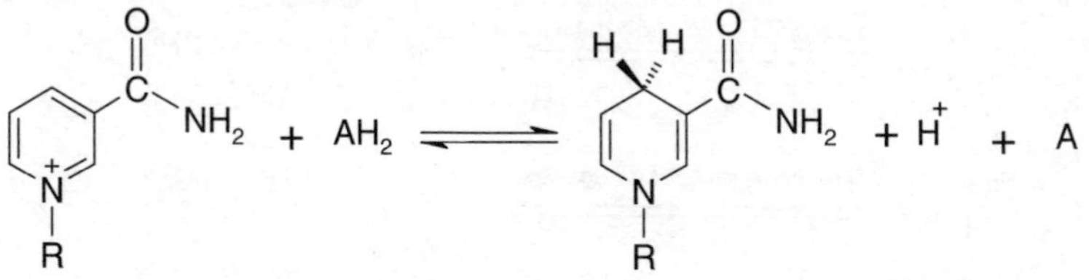

The reduced coenzymes are later reoxidised with the help of electron transport chain in the mitochondria accompanied with the formation of ATP (adenosine triphosphate), the energy carrier of the cell. The conversion of alcohol to aldehyde by *alcohol dehydrogenase* is a typical example of the reaction in which NAD⁺ acts as a coenzyme.

$$\text{Alcohol} \; + \; \text{NAD}^+ \; \rightleftharpoons \; \text{Aldehyde} \; + \; \text{NADH} \; + \; \text{H}^+$$

The reverse reaction i.e. the reduction of aldehyde to alcohol in the fermentation process is also catalysed by the same enzyme and is accompanied by the oxidation of NADH to NAD⁺. The proposed mechanism of the reaction involves the transfer of a hydride from the reduced coenzyme to the substrate.

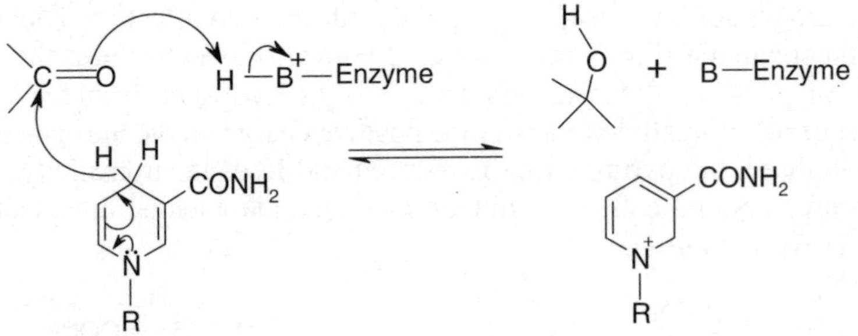

Nicotinamide adenine dinucleotide plays the role of coenzyme in many other enzymes. Some important reactions where NAD⁺ plays the role of a coenzyme are given below.

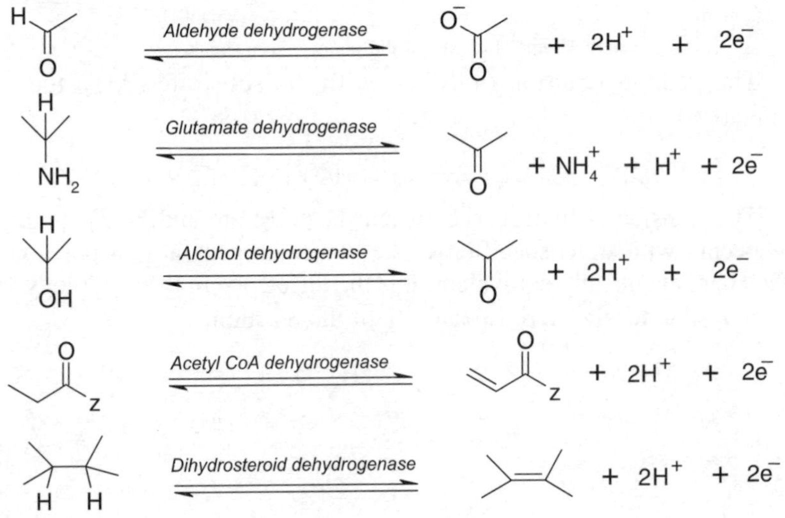

Since the hydrogen transfer reactions using NAD⁺ display a high degree of stereoselectivity they have been used extensively in organic synthesis. Chiral compounds requiring highly selective reagents to achieve enantiomeric purity can be made easily by using NADH and an appropriate enzyme, for example,

(97% ee)

Certain enzymes catalyse elimination, epimerisation or aldolisation reactions on such substrates which do not have requisite chemical groups to perform the said reaction. In such cases the coenzyme NAD⁺ performs a special function of transiently providing such chemical groups. The epimerisation of UDP-galactose to UDP-glucose, essential for the entry of galactose into the glycolytic pathway (related to the breakdown of glucose) or glyconeogenesis (synthesis of glycogen) in the cell is a typical example of this type.

4.7.2 UDP-Glucose

UDP-glucose (uridine diphosphate glucose) is a glucosyl donor for the biosynthesis of glycogen from glucose or other hexoses. It is an activated form of glucose like coenzyme A – an activated form of acetyl group or ATP which is an activated form of orthophosphate. UDP-glucose is formed by a reaction of glucose-1-phosphate with UTP catalysed by *UDP-glucose phosphorylase*. The reaction is reversible but is driven to the right by irreversible hydrolysis of the pyrophosphate to orthophosphate by an enzyme, *inorganic pyrophosphatase*. This hydrolysis continuously removes pyrophosphate the product of the first reaction, thereby driving it to the formation of UDP-glucose.

$$\text{Glucose-1-phosphate} + \text{UTP} \underset{\text{UDP-glucose phosphorylase}}{\rightleftharpoons} \text{UDP-glucose} + \text{PP}_i$$

$$\text{PP}_i + \text{H}_2\text{O} \xrightarrow{\text{inorganic pyrophosphatase}} 2\text{P}_i$$

_____

$$\text{Glucose-1-phosphate} + \text{UTP} + \text{H}_2\text{O} \longrightarrow \text{UDP-glucose} + 2\text{P}_i$$

This activated group of UDP-glucose can be transferred to the C-4 end of the elongating glycogen chain through a $\alpha(1 \rightarrow 4)$ linkage.

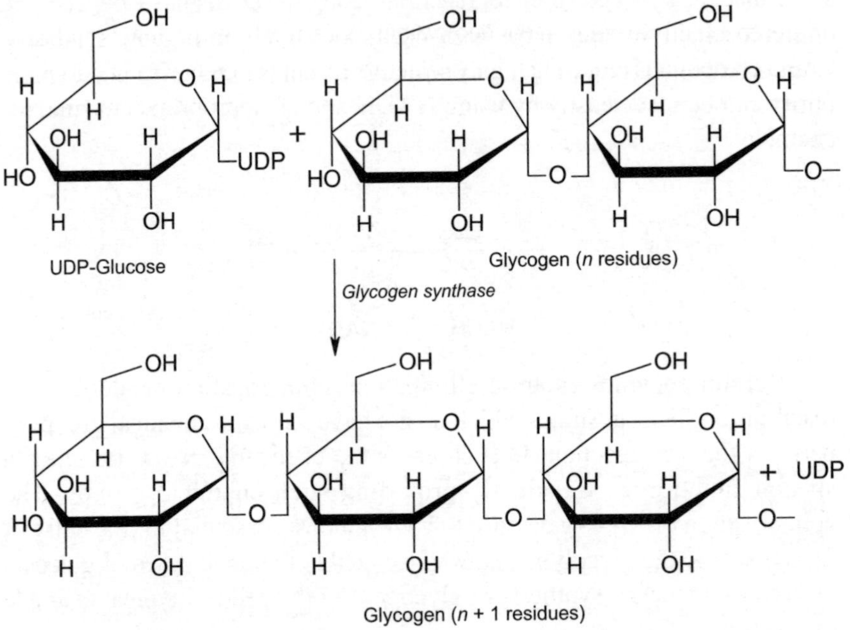

UDP-Glucose   Glycogen (*n* residues)

Glycogen synthase

Glycogen (*n* + 1 residues)

Since UDP-glucose can be obtained by the epimerisation of UDP-galactose, it plays another important role. It provides a means of storing excess galactose as glycogen in the body as UDP-glucose obtained can be stored as glycogen in the body.

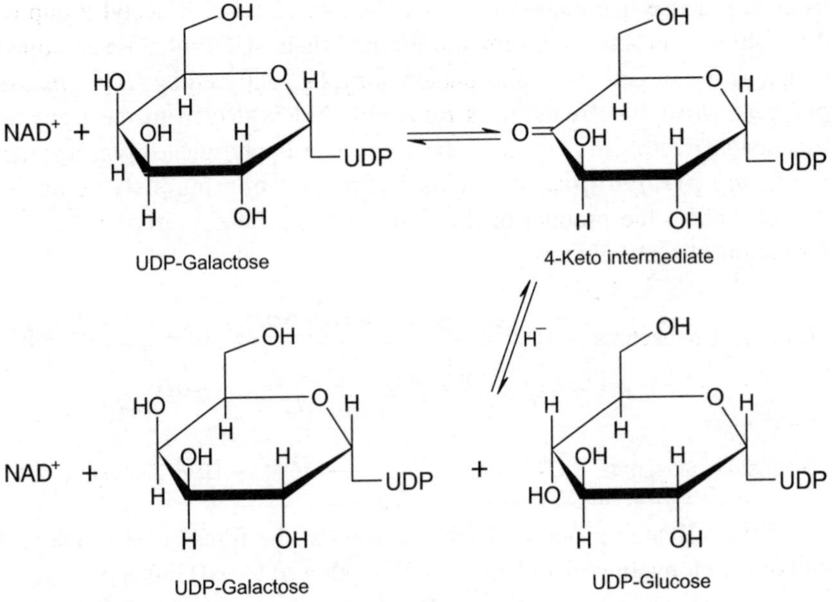

UDP-Galactose   4-Keto intermediate

UDP-Galactose   UDP-Glucose

The enzyme, *UDP-galactose-4-epimerase* associated with this reaction contains a tightly bound molecule of the coenzyme NAD⁺ or NADP. It has been demonstrated that NAD⁺ is transiently reduced to NADH in the reaction. Based on these observations it has been proposed that the oxidised coenzyme NAD⁺ accepts a proton from the – OH group at C-4 atom which gets transiently oxidised to a carbonyl group in the process.

In the next step the carbonyl group is reduced by accepting a hydride ion from NADH. Since this can happen from either side of the carbonyl group it generates UDP-galactose and its 4-epimer, UDP-glucose.

4.7.3 Flavin Mononucleotide (FMN) and Flavin Adenine Dinucleotide (FAD)

FMN and FAD are derived from **riboflavin** (also known as **vitamin B$_2$**) and act as coenzymes in electron transfer or oxidation-reduction reactions catalysed by certain enzymes. The word flavin has been derived from the Latin word *flavus* meaning yellow, the colour of the oxidised isoalloxazine ring present in the coenzyme. The enzymes that require FMN or FAD as cofactors are termed **flavoproteins**. These catalyse redox reactions acting as dehydrogenases, oxidases and monooxygenases (also called hydroxylases). Several flavoproteins also contain metal ions and are termed **metalloflavoproteins**. Flavoproteins function in different ways like, hydride transfer and substrate carbanion addition or by free radical mechanism. Therefore, these provide link between two electron and one electron transfer processes.

Structurally, FMN is formed by phosphorylation of riboflavin by the action of ATP with the help of enzyme *flavokinase.*

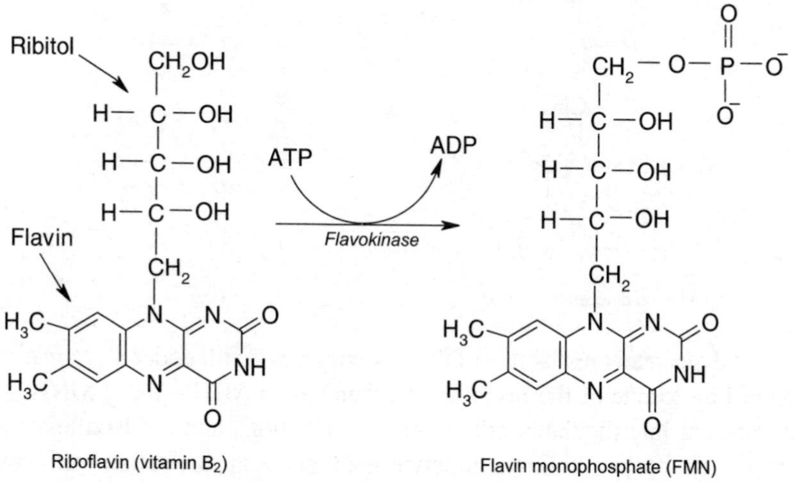

Riboflavin (vitamin B$_2$)   Flavin monophosphate (FMN)

Strictly speaking FMN is not a nucleotide, it is a derivative of ribitol –a pentahydroxy sugar while nucleotide contains ribose or deoxyribose sugar. FAD on the other hand, is synthesised by the reaction of FMN with ATP wherein the AMP moiety of ATP is transferred to FMN. The synthesis of FAD is controlled by enzyme flavin nucleotide pyrophosphorylase.

$$FMN \ + \ ATP \longrightarrow FAD \ + PP_i$$

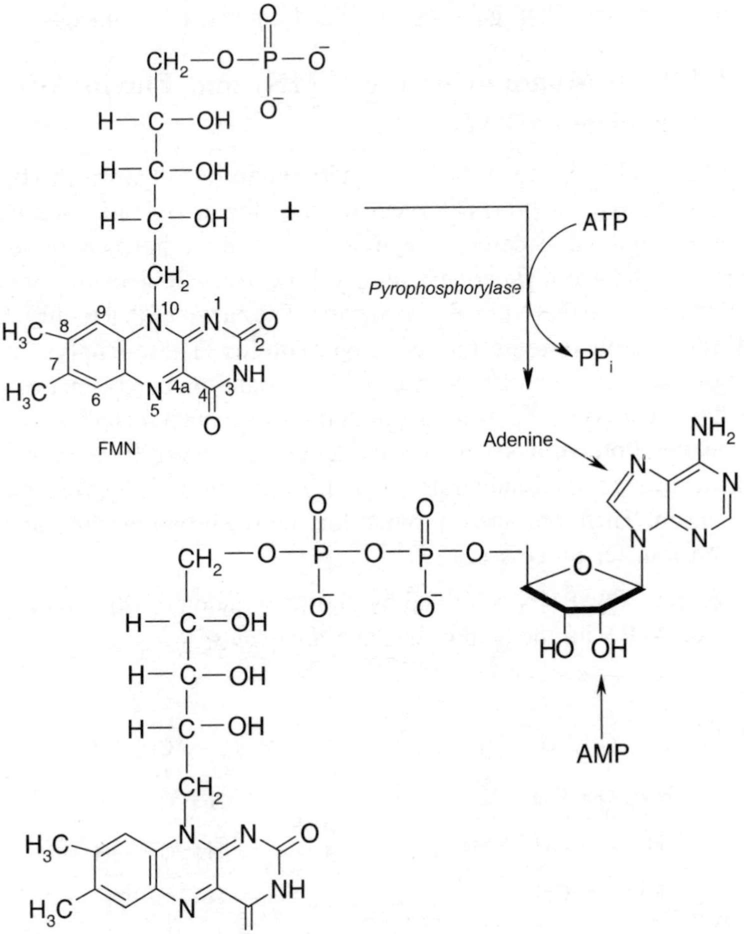

Flavin adenine dinucleotide (FAD)

The mechanism of action of these coenzymes is still under investigation. According to one of the proposed mechanisms $FADH_2$ and $FMNH_2$ act as transient intermediates in a number of reactions and the **isoalloxazine ring** system of the riboflavin moiety is the functional part of this coenzyme.

N-5, C-4a, N-10, N-1 and C-2 can be regarded as the catalytic entity; N-5 and C-4a positions being specifically important. The chromophore N=C–C=N plays the same role as played by nicotinamide ring in NAD⁺. Together, this structure operates in the transfer of hydrogens (or electrons).

This catalytic entity can exist in three spectrally distinguishable redox states. One of these is an oxidised form (yellow) another a reduced form (colourless) and there are two one-electron reduced forms (red and blue). The flavin coenzymes work as a switch between these one and two electron processes. Two-electron reduction of flavoquinone gives rise to flavohydroquinone while one-electron oxidation can give one of two semiquinone radicals. The red coloured 1H isomer is unstable while the blue coloured 5H isomer is quite stable and is of great importance in the one-electron transfer processes.

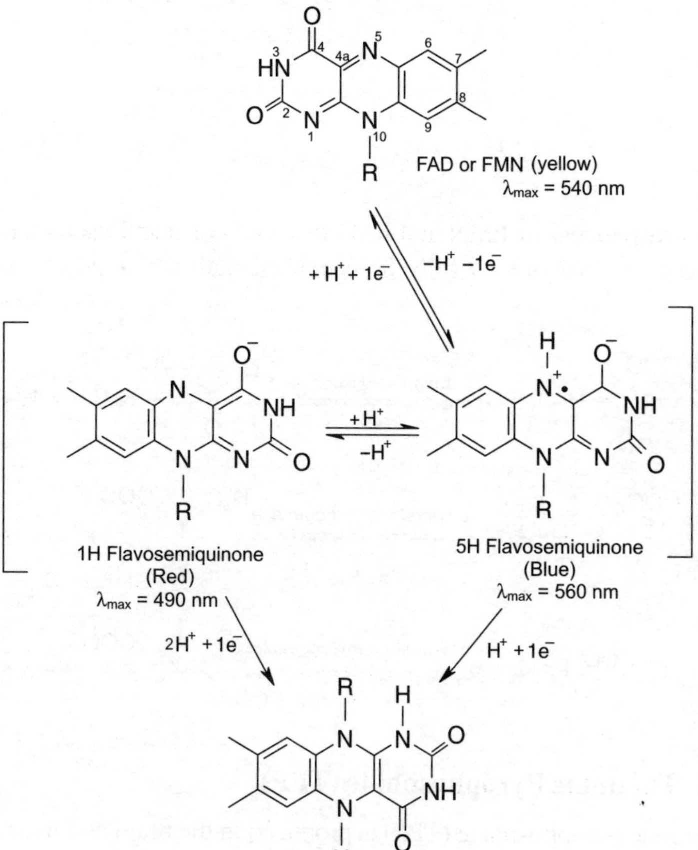

FAD or FMN (yellow)
λ_{max} = 540 nm

$+ H^+ + 1e^-$ $-H^+ - 1e^-$

$+H^+$
$-H^+$

1H Flavosemiquinone
(Red)
λ_{max} = 490 nm

5H Flavosemiquinone
(Blue)
λ_{max} = 560 nm

$2H^+ + 1e^-$ $H^+ + 1e^-$

FADH₂ or FMNH₂ (colourless)

Fatty acyl-CoA desaturase is an example of flavin-dependent enzyme, which catalyses an important step in the biosynthesis of unsaturated fats.

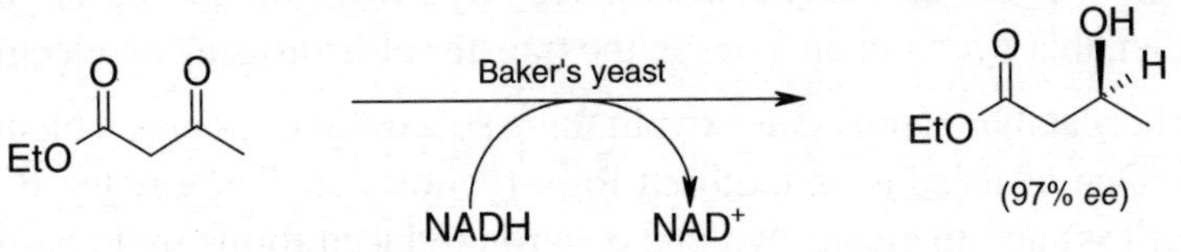

(97% ee)

The reaction is actually more complex than shown, and involves other cofactors, but FAD is the key cofactor for the enzyme. The most important aspect of the involvement of flavin coenzymes in hydrogen transfer is their ability to transfer hydrogen to oxygen as part of respiration process. In other words, they utilise molecular oxygen.

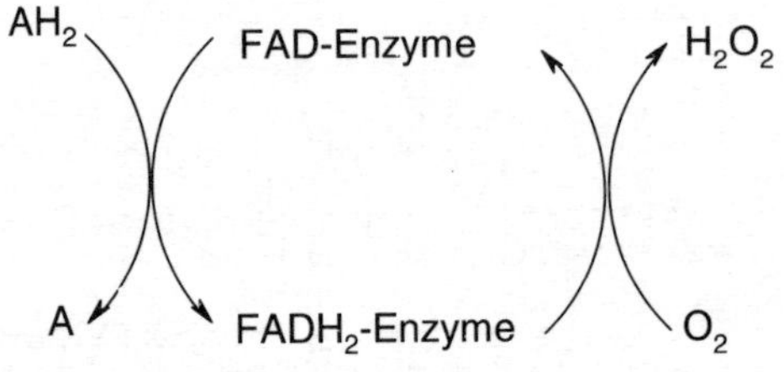

The involvement of FMN and FAD in variety of reactions on a range of substrates shows their versatility. Some representative examples are given below.

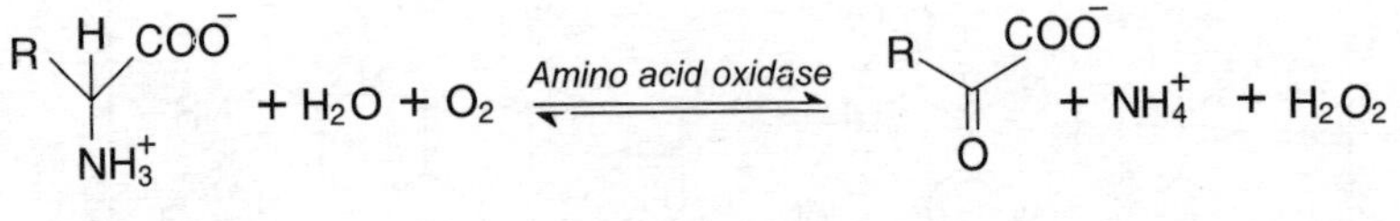

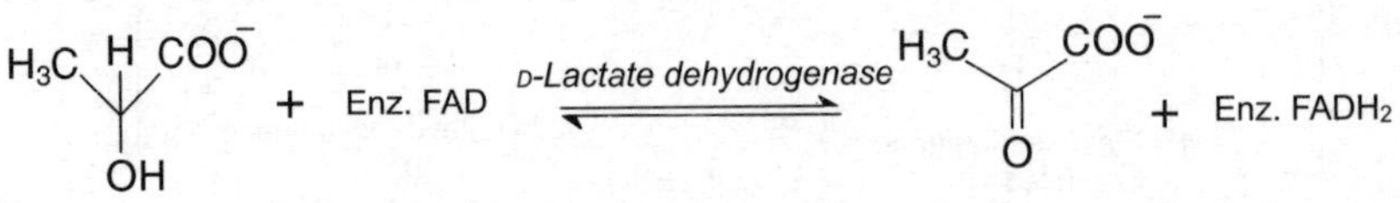

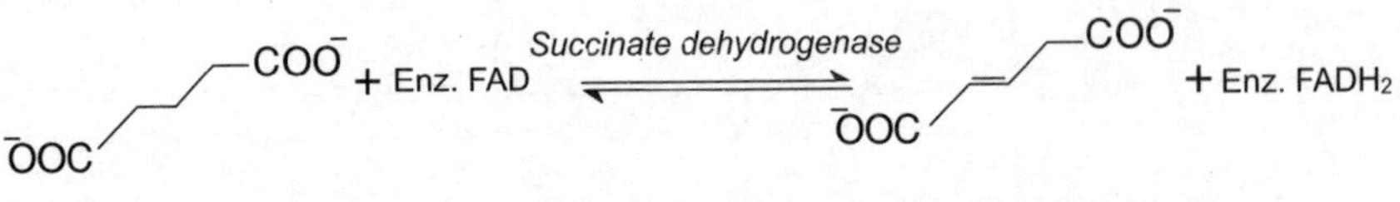

4.7.4   Thiamine Pyrophosphate (TPP)

Thiamine pyrophosphate (TPP) is produced in the brain and liver cells by phosphorylation of thiamine or vitamin B$_1$ with the help of the enzyme, *thiamine diphosphotransferase* or *TPP synthase*. In this process a pyrophosphate group is transferred from ATP to the vitamin and a molecule of AMP is generated as a by-product.

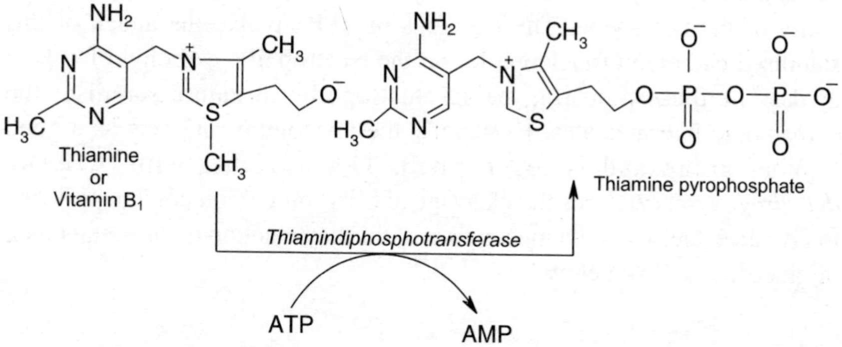

Thiamine
or
Vitamin B$_1$

Thiamindiphosphotransferase

ATP   AMP

Thiamine pyrophosphate

Thiamine, in turn consists of a substituted pyrimidine and a thiazole group coupled by a methylene bridge. TPP can stabilise carbanions and in doing so facilitates a number of crucial reactions in the cell.

TPP is necessary as a coenzyme for the enzymes catalysing decarboxylations of α-keto acids and the formation of C–C bonds between a carbanion stabilised compound on TPP and a carbonyl carbon of another compound. TPP catalyses the following types of reactions on α-keto acids produced by pyridoxal phosphate

- Oxidative carboxylation
- Transketolisation
- Nonoxidative decarboxylation
- Acetoin formation

The active portion of the TPP is thiazolium cation which can lose the acidic proton on the carbon between the S and the N atoms of the ring in a base catalysed mechanism to form a ylide.

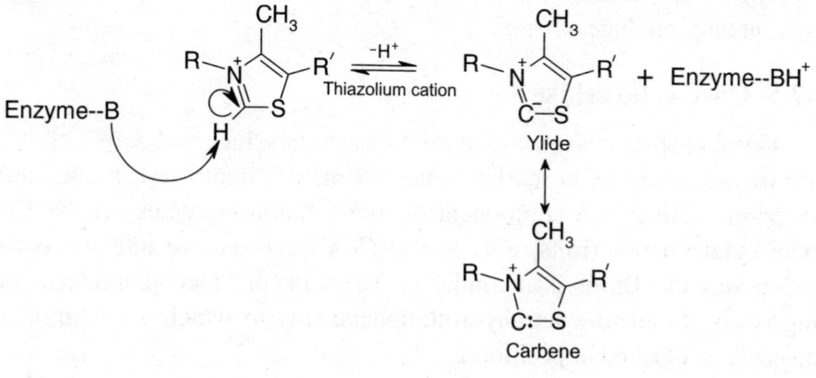

Enzyme--B

Thiazolium cation

Ylide

+ Enzyme--BH$^+$

Carbene

The negative charge of the carbanion is stabilised by the adjacent positive charge on the quaternary ammonium ion and the participation of *d*-orbitals of sulphur. The resonance stabilised carbanion so obtained is the reactive entity of the coenzyme. The reactions of TPP involve the attack of this stabilised carbanion (nucleophile) on the electrophilic carbonyl of α-keto acids. The thiazolium ring being electrophilic in nature stabilises the carbanionic intermediates. Eventually the thiazolium ring acts as a good leaving group and is regenerated. TPP, together with *pyruvate dehydrogenase*, catalyses the cleavage of a carbon-carbon bond in pyruvate to give acetaldehyde. The mechanism of this key reaction in the metabolism of glucose is shown below,

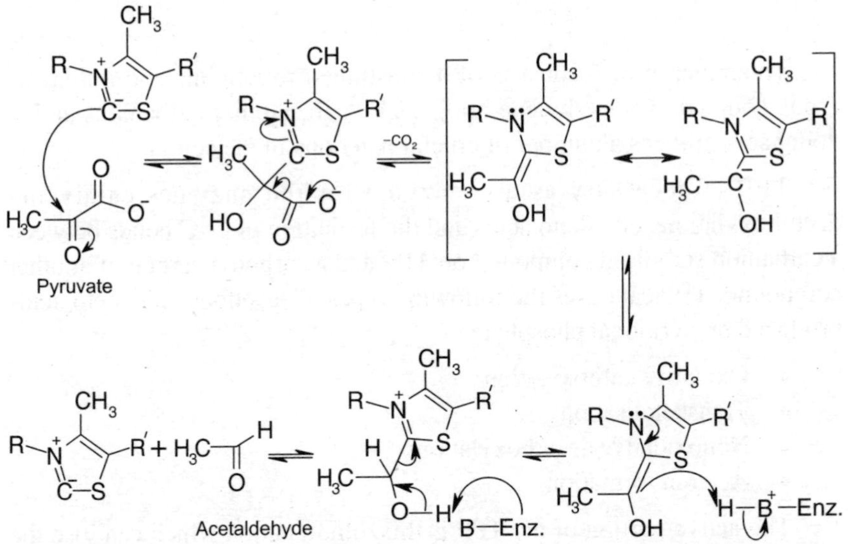

Since direct decarboxylation would give unstable carbonyl carbanion, it is unfavoured instead the enzyme stabilises the system to produce an aldehyde. This reaction occurs as part of the alcoholic fermentation of sugar to eventually produce ethanol.

4.7.5 Co-Carboxylase

Coenzymes that work with the enzymes catalysing transfer of carboxyl groups are called as co-carboxylases. **Biotin** (Vitamin H) is one such coenzyme acting as a component of multisubunit enzymes involved in carboxylation reactions, e.g. *acetyl-CoA carboxylase* and *pyruvate carboxylase* etc. Biotin is an imidazole derivative in which an imidazolone ring is *cis*- fused to a tetrahydrothiophene ring to which a valeric acid molecule is attached at position 2.

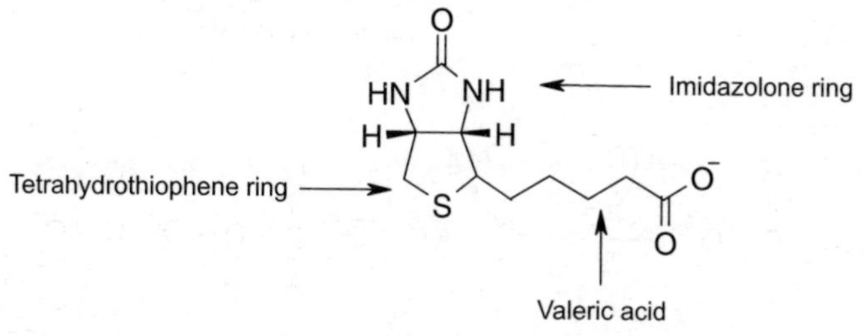

Biotin

The main function of biotin is to accept a carboxyl group and transfer it to a suitable substrate. All the steps of the reaction take place in the same multi-subunit complex. Biotin mediates two kinds of carboxylation reactions namely, direct carboxylations and trans carboxylations.

Direct Carboxylation

As the name indicates this involves direct carboxylation of the substrate. Conversion of acetyl-CoA to malonyl-CoA by *acetyl-CoA carboxylase* is a typical example. The multienzyme *acetyl-CoA carboxylase* contains three components, a biotin carboxyl carrying protein (BCCP), biotincarboxylase and carboxytransferase. In BCCP the carboxylic acid group of biotin is anchored to the ε-NH_2 group of a lysine residue on the enzyme via an amide bond.

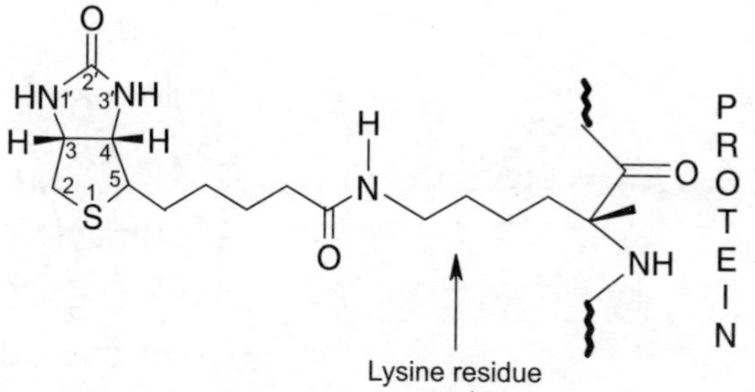

Biotin carboxyl carrying protein (BCCP)

In the first step $N^{1'}$ atom of the imidazolone ring, the seat of action, gets a carboxyl group from carbonic phosphoric anhydride to generate an activated intermediate called **carboxybiotin**. The carbonic-phosphoric anhydride (or carboxyphosphate) is formed by the action of ATP on bicarbonate ion and acts as a source of the carboxyl group. The addition of CO_2^- group from carboxyphosphate to $N^{1'}$ atom of biotin, is catalysed by *biotincarboxylase*.

$$ATP + HCO_3^- \rightarrow [PO_4-CO_2]^{3-} + ADP$$

Carboxyphosphate

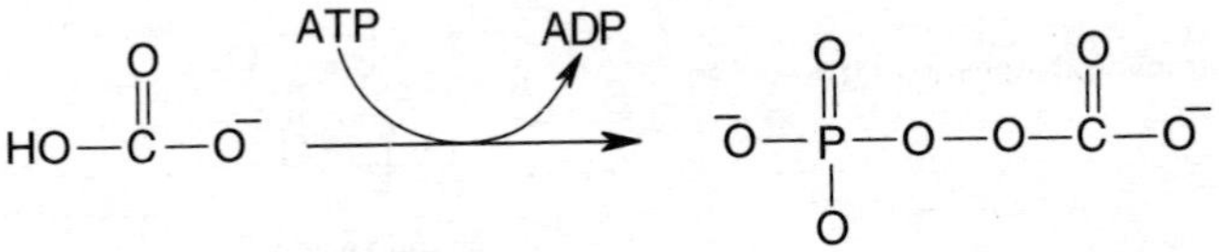

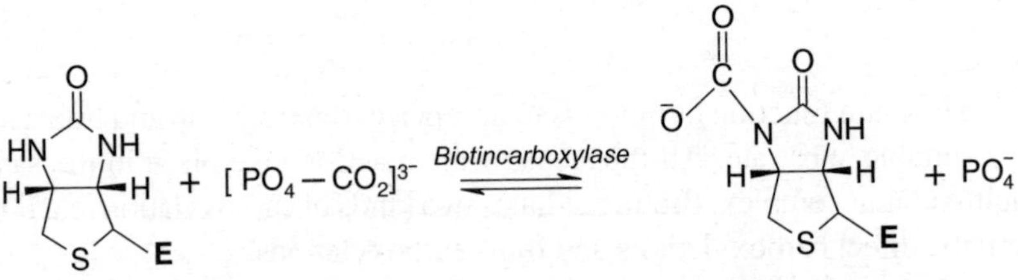

Biotincarboxylase

$N^{1'}$-Carboxybiotion

In the next step the activated carboxybiotin transfers the carboxylic acid group to acetyl-CoA with the help of the enzyme *carboxytransferase*. In the probable mechanism of the reaction, acetyl-CoA loses a proton to generate a carbanion that attacks the carboxyl group of the activated carboxybiotin followed by the removal of biotinate ion which picks up a proton to regenerate biotin attached to the protein.

These reactions can be summarised as

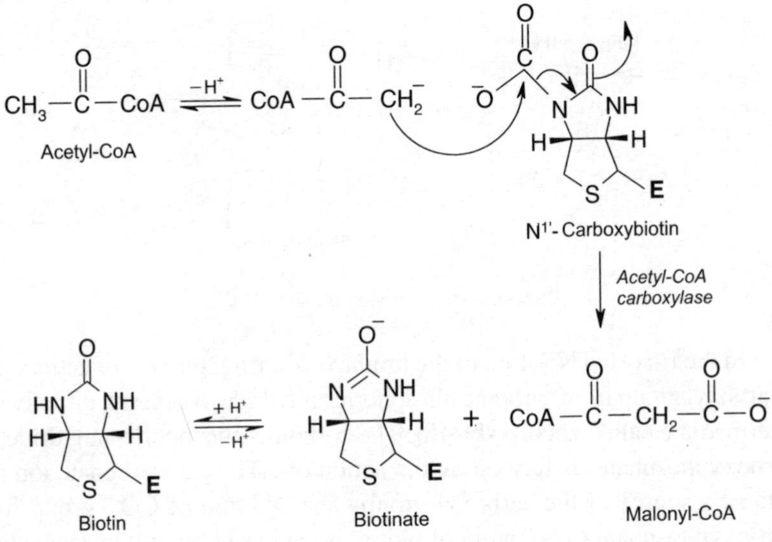

Acetyl-CoA

$N^{1'}$-Carboxybiotin

Acetyl-CoA carboxylase

Biotin   Biotinate   Malonyl-CoA

The generation of oxaloacetate from pyruvate also proceeds by a similar mechanism.

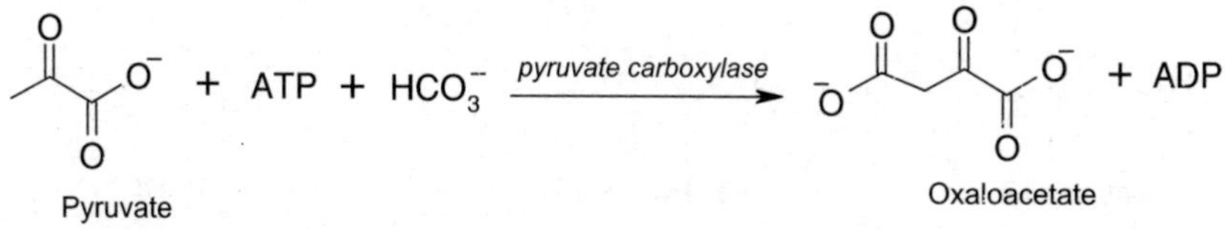

Pyruvate + ATP + HCO_3^- → (pyruvate carboxylase) → Oxaloacetate + ADP

Transcarboxylation

Transcarboxylation refers to the transfer of a carboxyl group from one species to the other e.g., methylmalonyl CoA transfers a carboxyl group to pyruvate to give oxaloacetate,

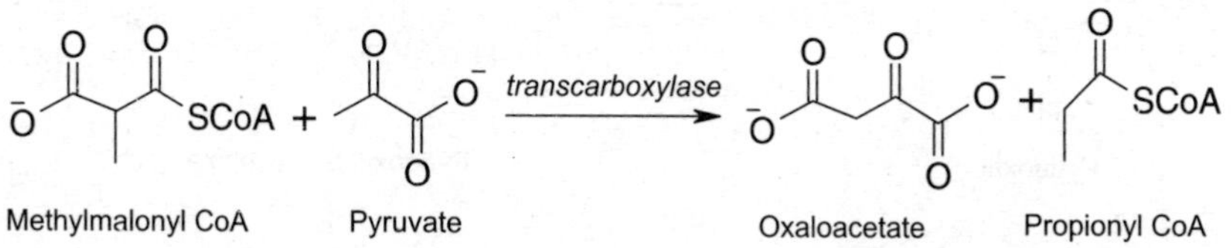

Methylmalonyl CoA + Pyruvate → (transcarboxylase) → Oxaloacetate + Propionyl CoA

Biotin is found in numerous foods and is also synthesised by intestinal bacteria and as such deficiencies of this vitamin are rare. Deficiencies are generally seen only after long antibiotic therapies which deplete the intestinal fauna or following excessive consumption of raw eggs. The latter is due to the affinity of the egg white protein, avidin, for biotin which prevents intestinal absorption of biotin.

4.7.6 Pyridoxal-5-Phosphate

Pyridoxal-5-phosphate is a versatile coenzyme engaged in a variety of reactions required for the synthesis and catabolism of α-amino acids. The reactions include transamination, racemisation, decarboxylation, α-β and β-γ eliminations etc. It is derived from vitamin B_6 which in fact is a group of three derivatives of pyridine namely, **pyridoxal, pyridoxamine** and **pyridoxine.** In diet these are present as pyridoxine, pyridoxal phosphate and pyridoxamine phosphate. All the three forms are important as these can be interconverted in the body. There may be some hydrolysis of the phosphate groups during digestion however, these are phosphorylated back by *pyridoxal kinase* present in the cell, with the help of ATP.

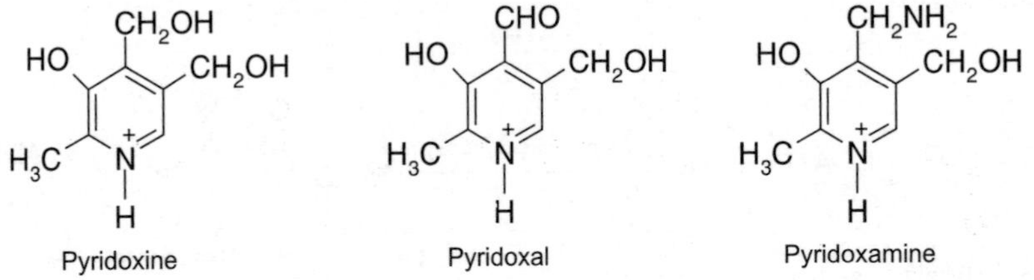

Pyridoxine   Pyridoxal   Pyridoxamine

Of the three constituents of vitamin B_6, it is pyridoxal-5-phosphate (PLP) that is active as coenzyme.

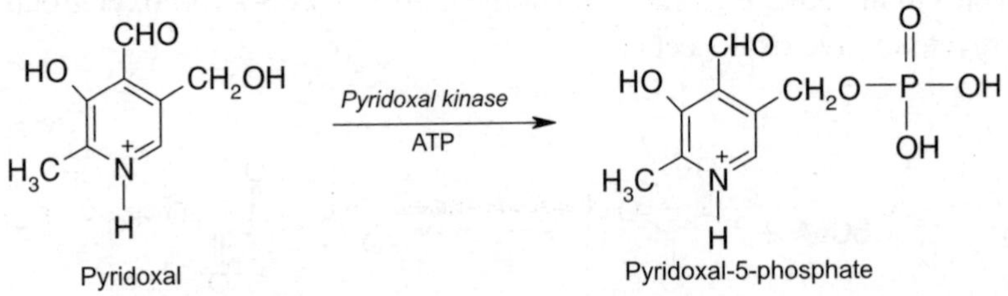

Pyridoxal   Pyridoxal-5-phosphate

Pyridoxal phosphate acts by forming an **aldimine** or **Schiff's base** adduct with the amino group of the α-amino acid and catalyses bond cleavages by stabilising the electronic environment of the α- and β- carbons of the adduct intermediates. The adduct can undergo different changes depending on the enzyme involved and cause different overall changes. Similar to TPP, loss of a H as a proton on the α- carbon yields a carbanion that is resonance stabilised by bond shifts in the adduct. When β-carbon has -OH, -SH or phosphate as functional group, pyridoxal phosphate can act to effect β-elimination and so on.

The role of pyridoxal phosphate in transamination reaction in the metabolism of α-amino acids is quite well studied and interesting. Transamination refers to the transfer of an α-amino group from an α- amino acid to a α-keto acid to give a new amino acid and a different α-keto acid. These reactions are catalysed by aminotransferases also called transaminases. More than 50 aminotransferases are known and all need pyridoxal phosphate as coenzyme. One of the most important aminotransferases is *aspartate aminotransferase* which catalyses the transfer of α- amino group of aspartate to α-ketoglutarate producing oxaloacetate and glutamate. The reaction can be represented as follows.

Aspartate + α – Ketoglutarate ⟶ Oxaloacetate + Glutamate

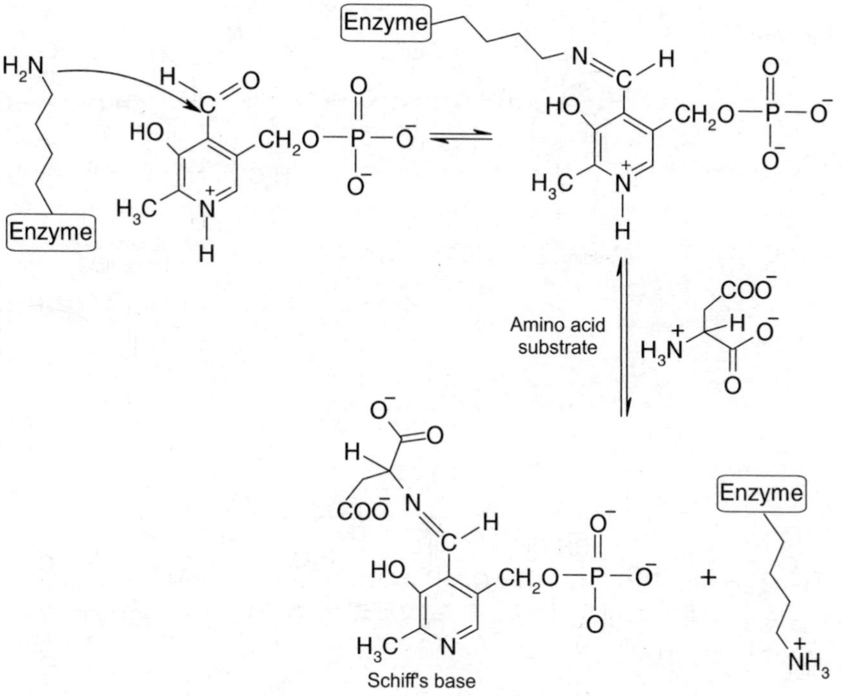

Schiff's base

Transamination reaction is proposed to have the following mechanism. The coenzyme, pyridoxal phosphate, is initially bound to the ε- amino group of a lysyl residue of the enzyme.

The incoming substrate forms a schiff's base with pyridoxal phosphate bound to lysyl residue, by displacing the ε-NH_2 group and releasing the lysyl residue of the enzyme. This intermediate (schiff's base) called an aldimine contains a double bond between the α-N of the amino acid and the carbon of the coenzyme. The proton attached to the α- carbon atom is lost to give rise to a quinonoid structure which on reprotonation forms a ketimine containing unsaturation between Cα-N bond. This ketimine on hydrolysis yields an α- keto acid and pyridoxamine phosphate.

Effectively, the net result of these steps amounts to the transfer of α-NH_2 group of the amino acid to the coenzyme and releasing α- keto acid (oxaloacetate).

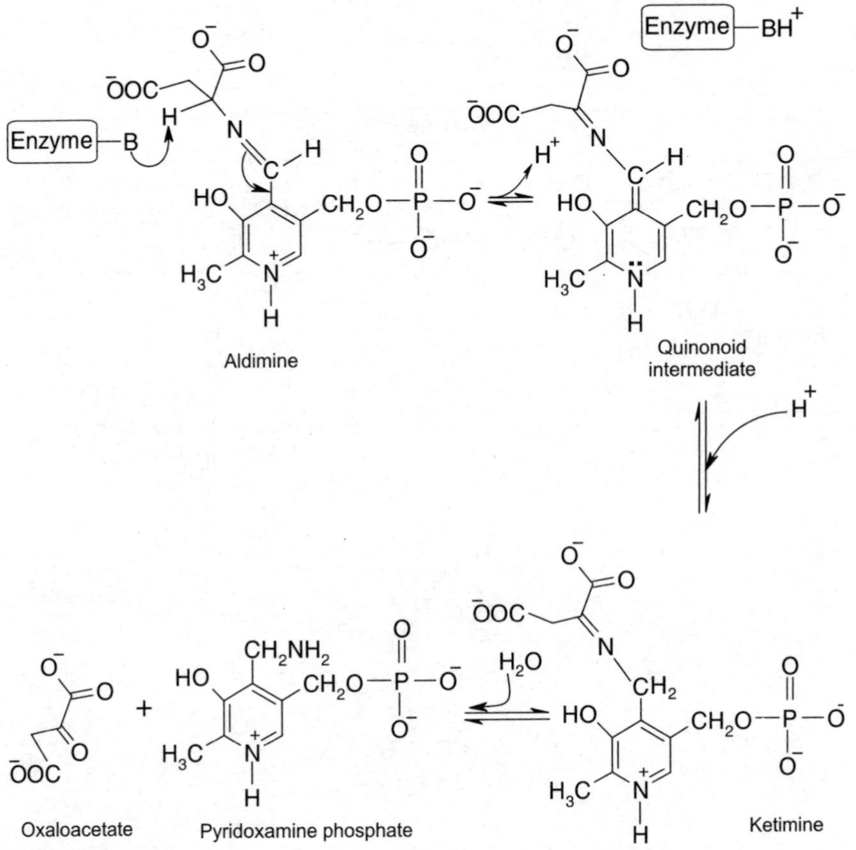

Aldimine

Quinonoid
intermediate

Oxaloacetate   Pyridoxamine phosphate   Ketimine

In the next set of reactions the amino group is transferred from the coenzyme to another α-keto acid. The enzyme-pyridoxamine complex now binds to α- ketoglutarate and hands over the above mentioned NH_2 group to α-ketoglutarate yielding back pyridoxamine phosphate-enzyme complex. To execute this, pyridoxamine phosphate condenses with the α-keto group of α-ketoglutarate to give an adduct. A proton from C_4' atom is lost to yield a quinonoid structure. Accepting a H at the C_α facilitates the release of glutamate with a concomitant regeneration of enzyme- pyridoxal phosphate complex.

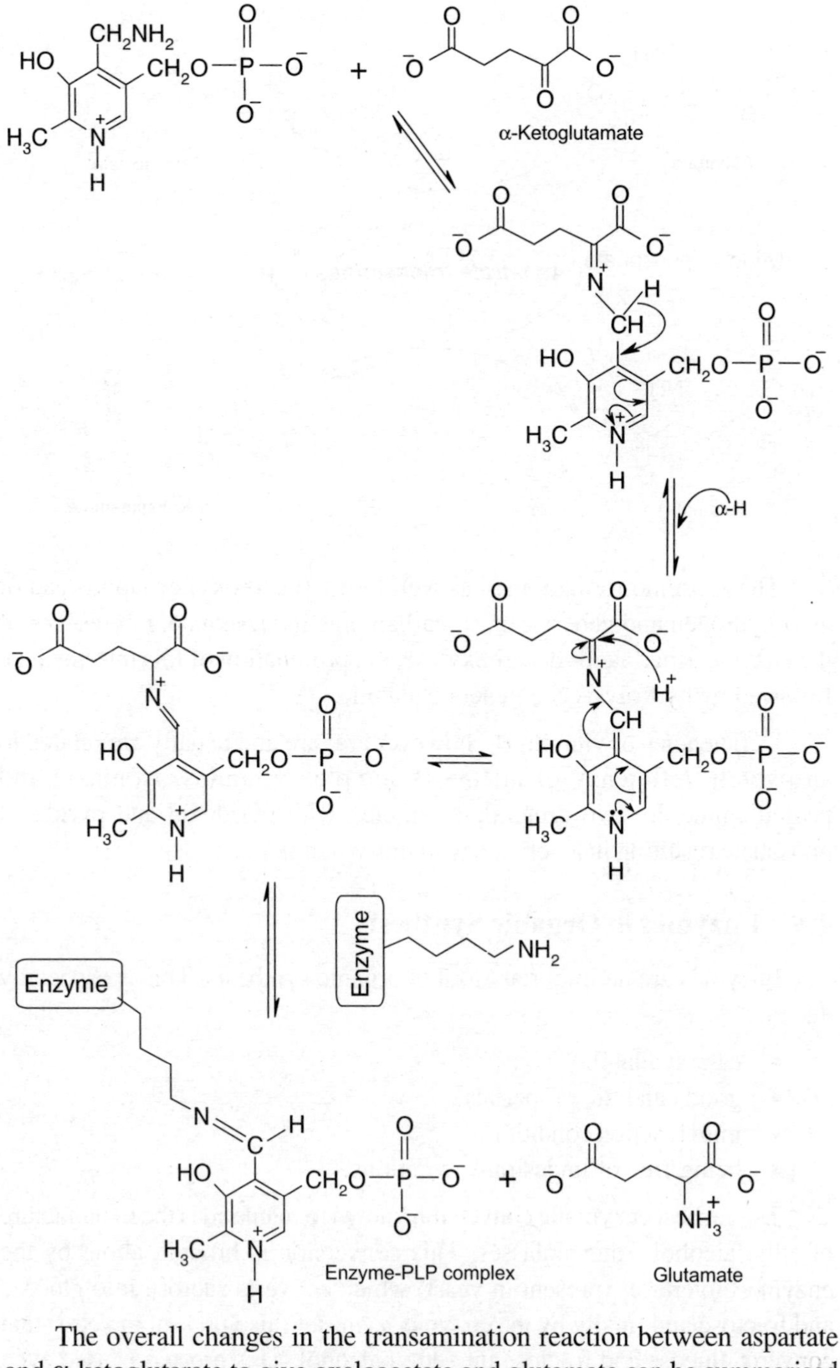

The overall changes in the transamination reaction between aspartate and α-ketoglutarate to give oxaloacetate and glutamate can be summarised as

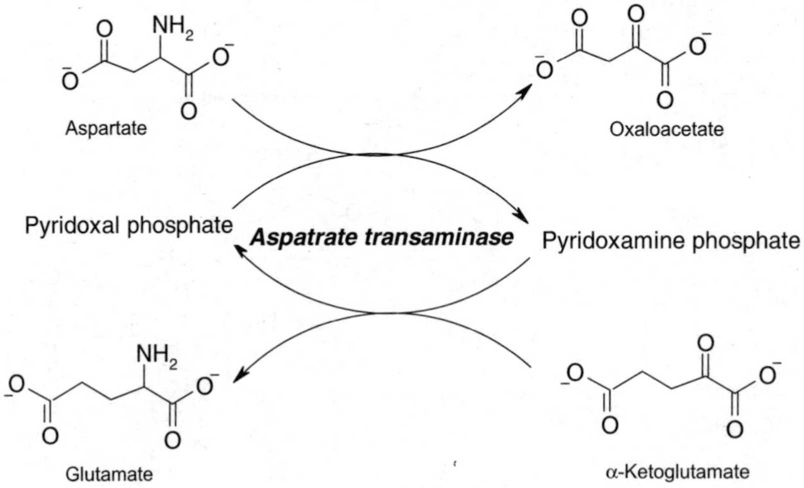

The ketimine formed, may as well lose a β carboxyl group instead of an α hydrogen and generate an α- carbanionic intermediate. In presence of the enzyme aspartate β- decarboxylase, the protonation of this intermediate followed by hydrolysis can generate alanine.

Deficiencies of vitamin B_6, however are rare and usually are related to an overall deficiency of all the B-complex vitamins. Isoniazid and penicillamine are two drugs that complex with pyridoxal and pyridoxal phosphate resulting in a deficiency in this vitamin.

4.8   Enzymes in Organic Synthesis

Enzymes are an important tool in organic synthesis. This is especially due to:

- easy availability
- good catalytic properties
- mild reaction conditions
- being free of undesirable reactions

The earliest enzymatic conversion known to mankind is the manufacture of ethyl alcohol from molasses. This conversion is brought about by the enzyme 'invertase' (present in yeast) which converts sucrose into glucose and fructose and finally by the enzyme 'zymase' (also present in yeast) that converts glucose and fructose into ethyl alcohol. This process is used even today for the manufacture of ethyl alcohol. The reactions are as given below.

$$C_{12}H_{22}O_{11} \xrightarrow{\text{invertase (in yeast)}} C_6H_{12}O_6 \;+\; C_6H_{12}O_6$$

Sucrose   Glucose   Fructose

$$C_6H_{12}O_6 \xrightarrow{\text{zymase (in yeast)}} 2\,CH_3CH_2OH \;+\; 2\,CO_2$$

Glucose and Fructose   Ethyl alcohol

Two other conversions, known since early times are the enzymatic conversion of ethyl alcohol into acetic acid by bacterium acetic in presence of air (process is known as quick-vinegar process) and the conversion of lactose into lactic acid by *Bacillus acidic lactic.*

$$CH_3CH_2OH \;+\; O_2 \xrightarrow{\text{Bacterium acetic}} CH_3COOH \;+\; H_2O$$

Ethyl alcohol   Acetic acid

$$C_{12}H_{22}O_{11} + H_2O \xrightarrow{\text{Bacillus acetic lactic}} 4\,CH_3CH(OH)COOH$$

Lactose   Lactic acid

The enzymatic transformations given above are referred to as **fermentations.** These and some of the other earlier applications used organisms as such for the transformations. Due to the associated messy procedures these processes did not find many uses. Once the isolation of enzymes from the organisms became possible at good scale, these were used in numerous industrial and synthetic processes. Recent advances in the area of genetic engineering have facilitated the production of large quantities of purified enzymes which in turn has increased the scope of industrial and synthetic applications of enzymes.

The enzymatic conversions or transformations have definite advantages, some of these are:

- Generally the reactions are performed in aqueous medium and at ambient pH, temperature and pressure.
- The reactions involve only one step and are much faster than non-enzymatic reactions.
- The conversions are regio, stereo and chemospecific

A significant advantage of enzymatic reaction is that same conversions or transformations, which are not possible by conventional chemical means can be achieved by enzymes. Some examples of this type are given below.

Electrophilic substitution next to nitrogen in heterocycle

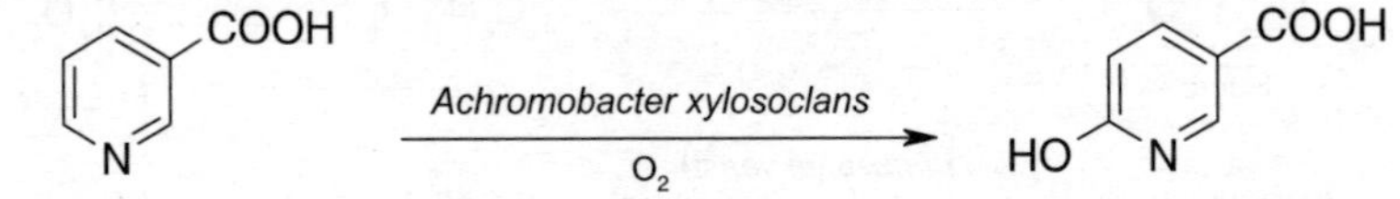

Oxidation of hydrocarbon in the side chain of heterocycle to carboxylic acid.

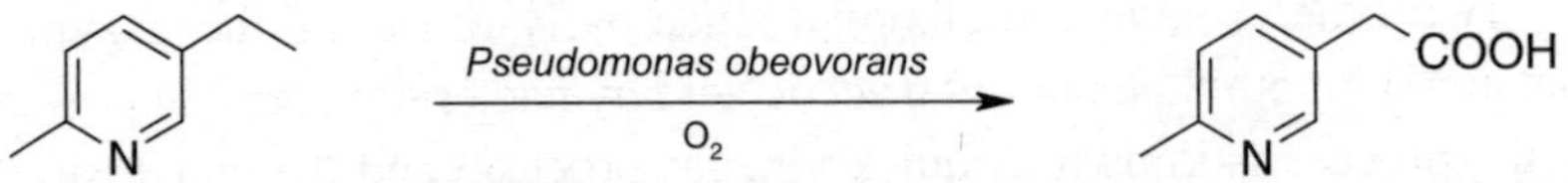

The most important application of enzymes in organic synthesis involves enzymatic oxidations, hydroxylations, hydrolysis, reductions and isomerisations. A brief discussion on these is given below.

4.8.1 Enzymatic Oxidations

Considerable amount of work has been reported on enzymatic oxidations of aromatic nucleus. Benzene and substituted benzenes give corresponding cis-diols on oxidation with *dioxygenases* from *Pseudomonas putida* in the presence of oxygen.

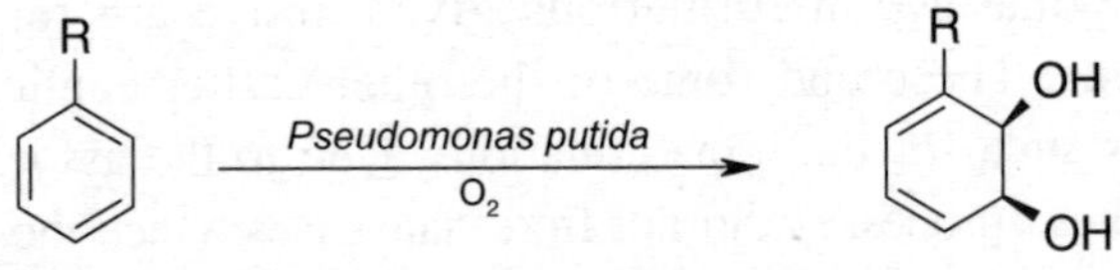

benzene	R= H
toluene	R= CH$_3$
chlorobenzene	R= Cl
styrene	R= -CH=CH$_2$
phenylacetylene	R= —C≡CH

The products of the reaction are often obtained in optically pure form and their derivatives (e.g., acetonides) may act as substrate for various cycloadditions to give important molecules. For example, the acetonide of the cyclohexa-3, 5-diene-1, 2-cis-diol obtained from the oxidation of benzene in presence of *Pseudomonas putida* can undergo Diels-Alder [4+2] addition to give tricyclic compounds.

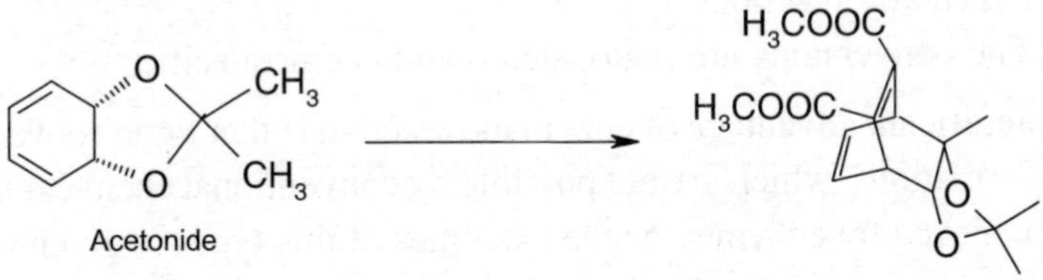

Acetonide

The oxidation of hydrocarbons is an important transformation whose product is used as feedstock in chemical industries. Generally it is possible under harsh conditions only, however, with the help of enzymes these oxidations can be performed under mild conditions. For example, oxidation of *p*-xylene to terephthalic acid by chloroperoxidase enzyme extracted from *caldariomyces fumago*.

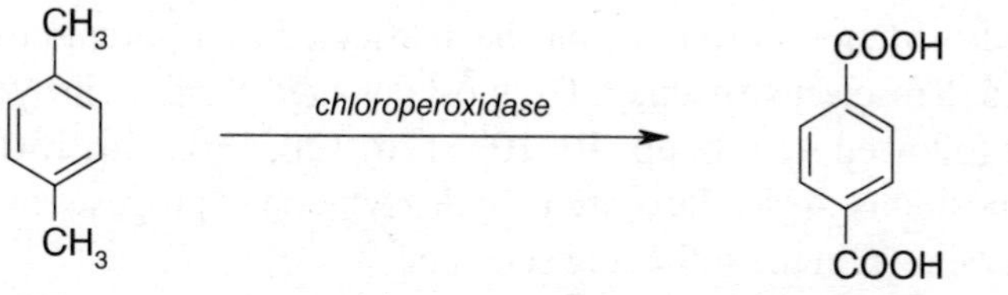

p-Xylene   Terephthalic acid

Secondary alcohols on enzymatic oxidation give the corresponding ketones. A secondary alcohol group can be oxidised to the corresponding ketone even in the presence of primary alcoholic group as shown below.

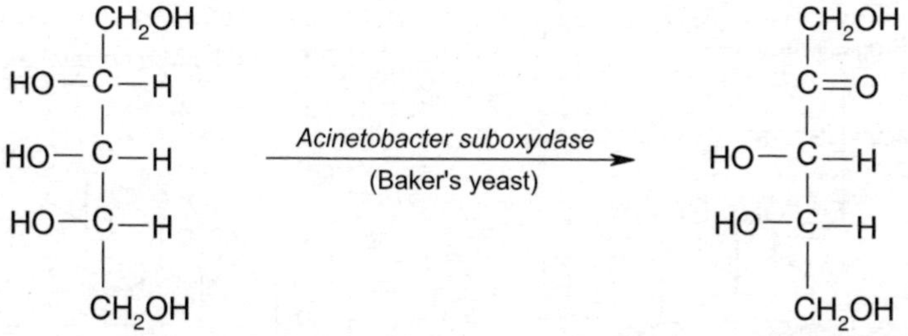

Baeyer-villiger oxidations can also be conveniently performed using enzymes. The monooxygenases from *Acinetobacter* are used in association with its cofactor NADPH. The following are two such examples:

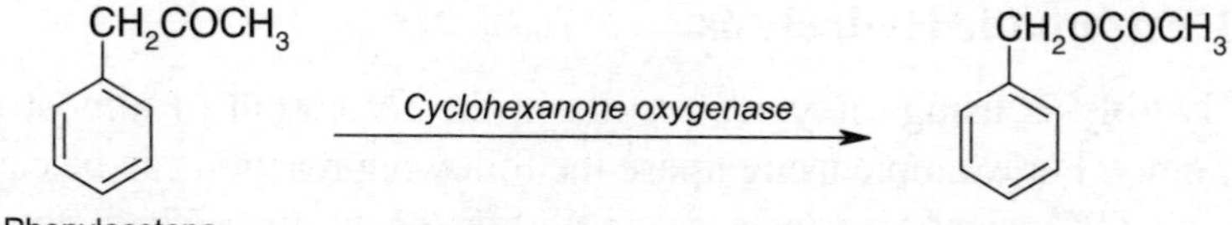

Phenylacetone   Benzyl acetate

Ring expansion of cyclic ketones by microorganisms containing an oxidative flavin based enzyme is used to generate chiral lactones. For example:

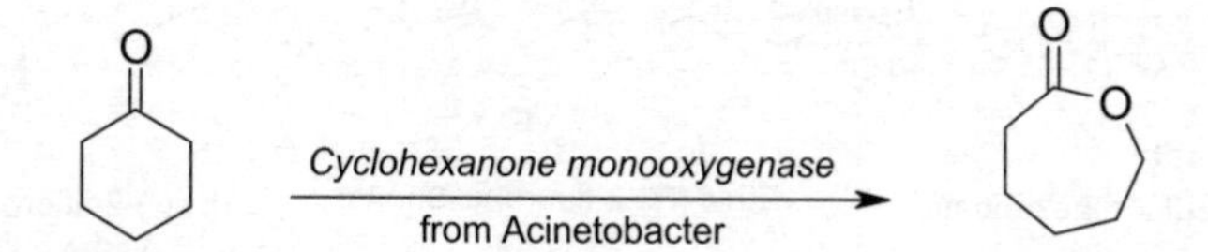

Cyclohexanone   ε-Caprolactum

Amino groups can be oxidized into nitro group by enzymes.

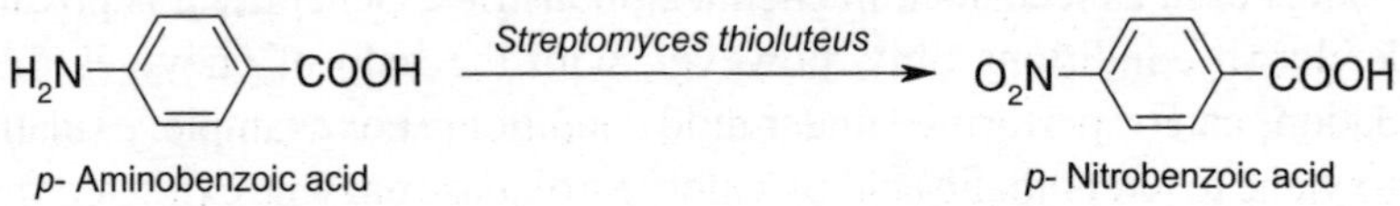

p- Aminobenzoic acid   p- Nitrobenzoic acid

4.8.2 Enzymatic Hydroxylation

Hydroxylation in steroids has been affected in different positions in a regio and steroselective manner. The most favoured positions in hydroxylation are 11α followed by 11β, 6β, 7α, 10β, 12α, 12β, 13β, 14α, 15α, 15β, 17α and 7β positions. As an illustration, hydroxylation of progesterone gives the products shown in the following scheme.

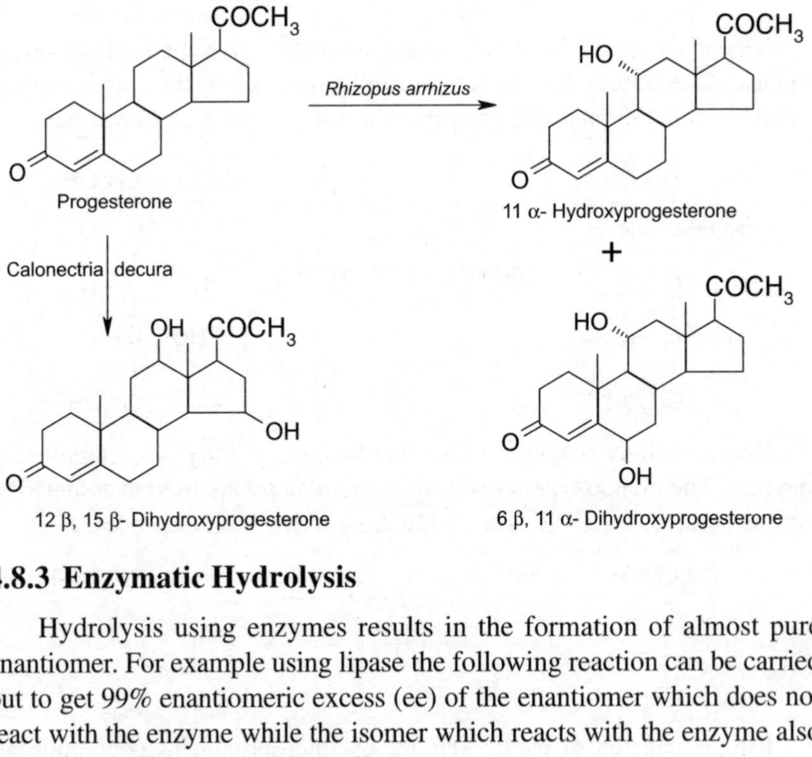

Progesterone

11 α- Hydroxyprogesterone

+

12 β, 15 β- Dihydroxyprogesterone

6 β, 11 α- Dihydroxyprogesterone

4.8.3 Enzymatic Hydrolysis

Hydrolysis using enzymes results in the formation of almost pure enantiomer. For example using lipase the following reaction can be carried out to get 99% enantiomeric excess (ee) of the enantiomer which does not react with the enzyme while the isomer which reacts with the enzyme also gives the hydrolysed product in 69% ee.

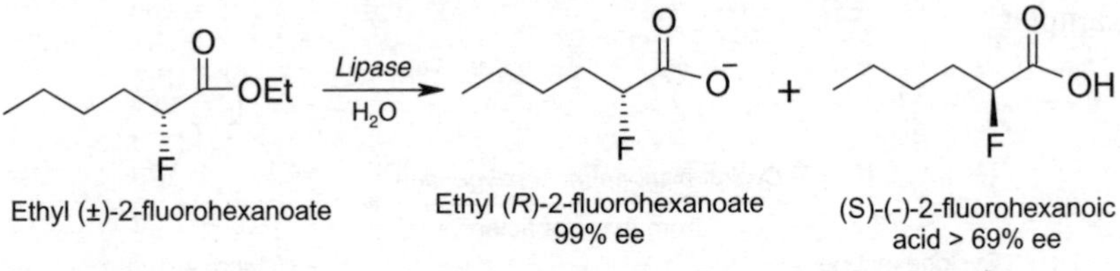

Ethyl (±)-2-fluorohexanoate

Ethyl (R)-2-fluorohexanoate
99% ee

(S)-(-)-2-fluorohexanoic
acid > 69% ee

Enzymes have been successfully used for selective hydrolysis as shown in the examples given below.

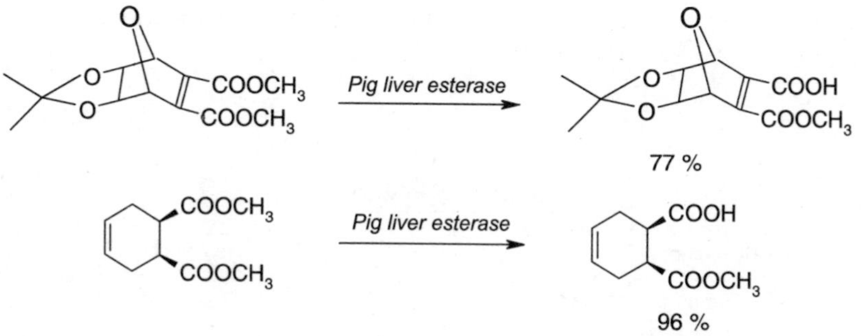

77 %

96 %

Enantioselective hydrolysis of the following has been affected by hog pancreatic lipase.

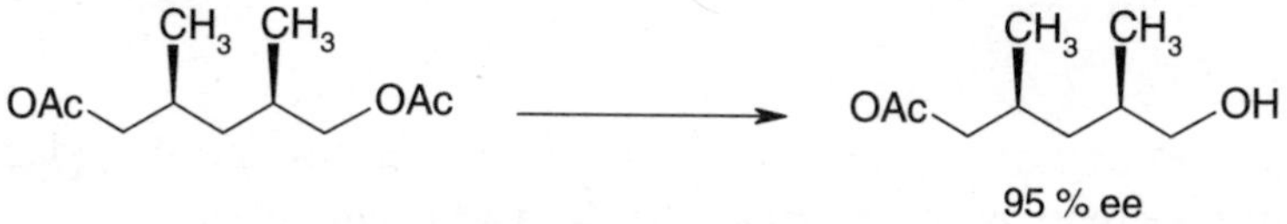

95 % ee

Selective hydrolysis of nitrile group can be carried out with the help of the enzyme *nitrile hydratase* under very mild conditions.

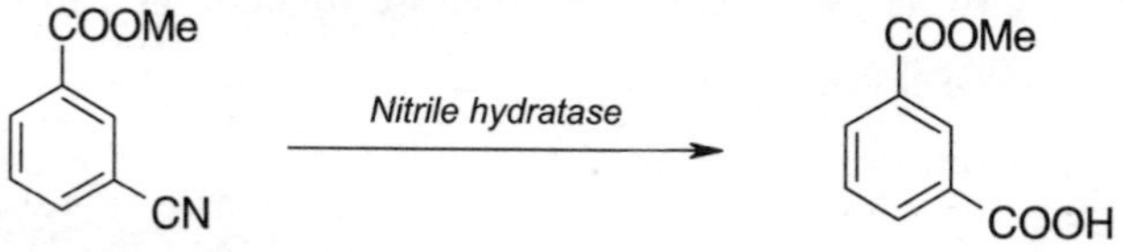

4.8.4 Enzymatic Reductions

The organisms like bakers' yeast and isolated enzymes have been extensively used for the reduction of carbonyl groups to the corresponding alcohols. Bakers' yeast has been used to reduce a wide range of simple ketones, ß-ketoesters and cyclic diketones, etc. Yeast alcohol dehydrogenase (YAD) and horse liver alcohol dehydrogenase (HLADH) are commonly employed for reductions. In the reductions, change of enantioselectivity is observed with the change of substrates.

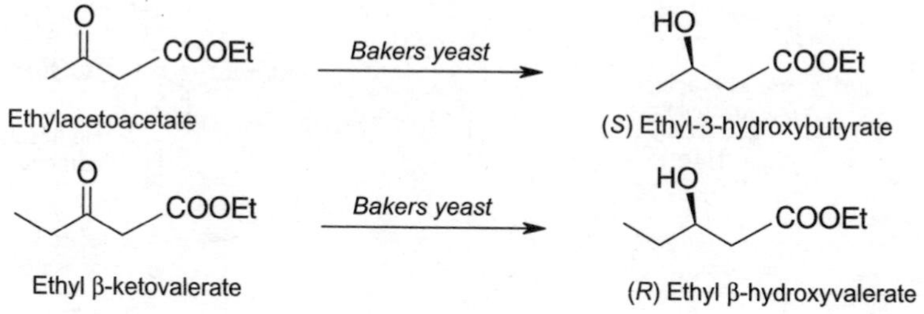

Ethylacetoacetate

(S) Ethyl-3-hydroxybutyrate

Ethyl β-ketovalerate

(R) Ethyl β-hydroxyvalerate

The selectivity of the reactions as given above was inconsistent with the predictions based on Prelog's rule. Enzymatic reduction of 2-butanone gave (R)-alcohol but reduction of 2-hexanone gave (S)-alcohol as shown below.

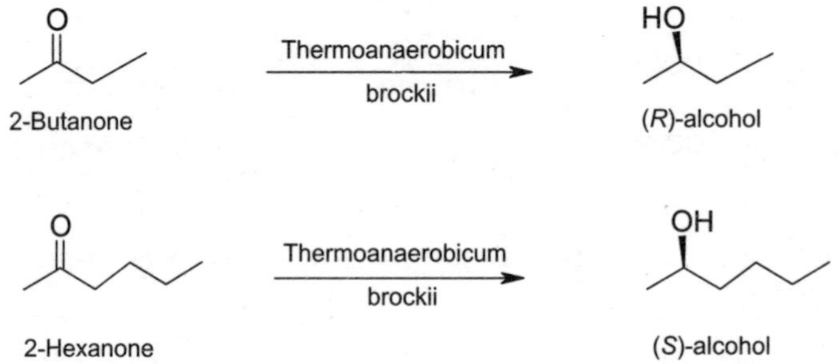

2-Butanone → Thermoanaerobicum brockii → (R)-alcohol

2-Hexanone → Thermoanaerobicum brockii → (S)-alcohol

The above examples illustrate that the enantioselectivity of the reductions and its selectivity is dependant on the size and nature of the group around the carbonyl group.

Bakers' yeast has also been used to cause reduction of aromatic nitro groups to the corresponding amines.

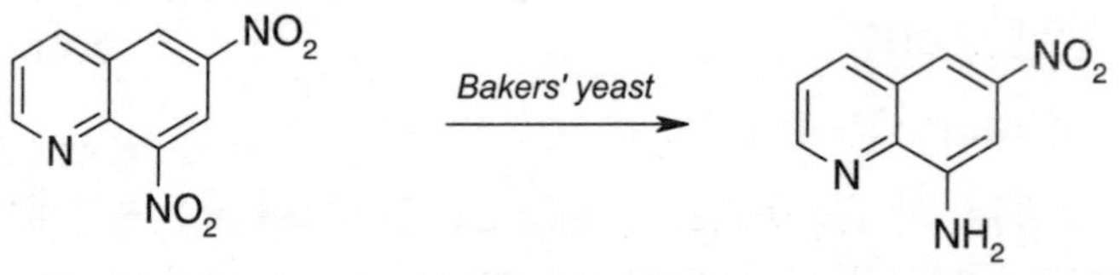

The reduction of cyclohexanone to cyclohexanol using horse liver alcohol dehydrogenase is another example of enzymatic reduction. This requires the presence of a stoichiometric amount of the reducing cofactor (NADH) which is provided by simultaneously oxidising a sacrificial substrate like Hantzsch carboxylate that regenerates NADH from NAD^+.

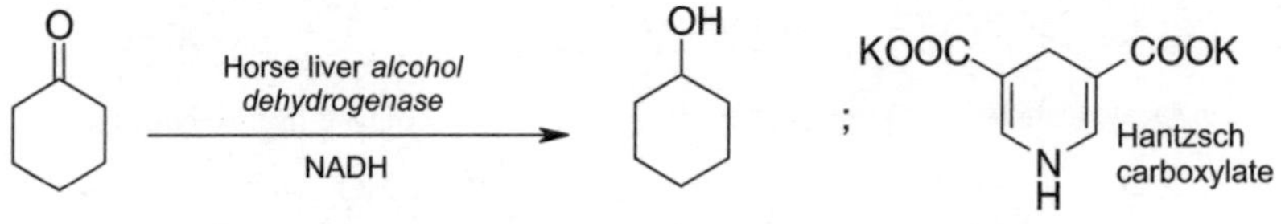

Horse liver *alcohol dehydrogenase* / NADH

Hantzsch carboxylate

4.8.5 Enzymatic Isomerisations

Only few examples of enzymatic isomerisation have been recorded. The most important is the production of high-fructose corn syrup from glucose using glucose isomerase.

4.8.6 Pharmaceutical Applications of Enzymes

One most common example is the enzymatic conversion of penicillin into 6 aminopenicillanic acid (6APA) by the enzyme '*penacylase*'. The hydrolytic product, 6APA is obtained in one step on a large scale. The product is free of any impurities and is used to prepare semisynthetic penicillins. The chemical conversion, however, requires a number of steps as follows.

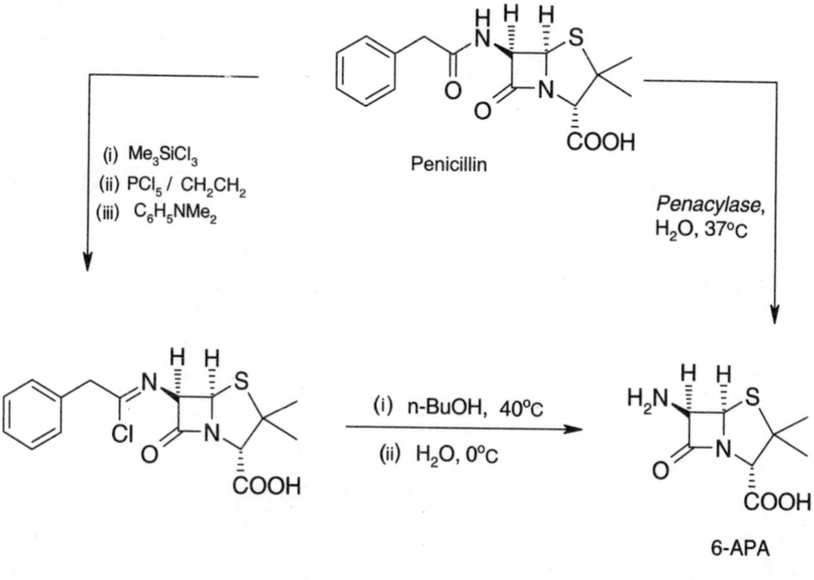

Penicillin

(i) Me$_3$SiCl$_3$
(ii) PCl$_5$ / CH$_2$CH$_2$
(iii) C$_6$H$_5$NMe$_2$

Penacylase,
H$_2$O, 37°C

(i) n-BuOH, 40°C

(ii) H$_2$O, 0°C

6-APA

Another well known example of enzymatic conversion is the conversion of Reichsteins compound into cortisol by the enzyme '*11β-hydroxylase*' into cortisol and finally conversion of cortisol into prednisolone with enzyme $\Delta^{1,2}$-*dehydrogenase*'.

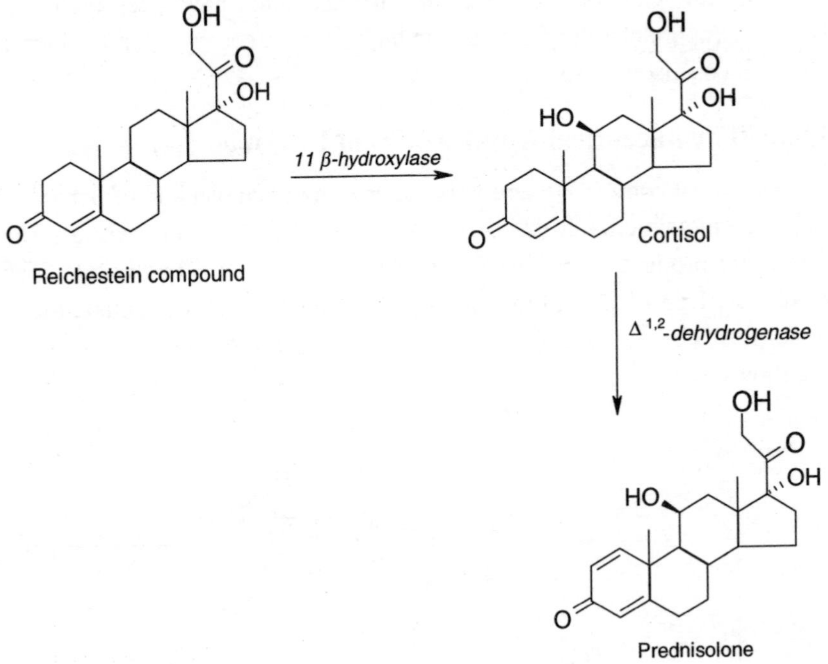

Reichestein compound

11 β-hydroxylase

Cortisol

$\Delta^{1,2}$-dehydrogenase

Prednisolone

The above enzymatic conversions are rapid, regio- and stereoselective.

EXERCISES

1. What are enzymes? List important characteristics of enzymes.

2. Briefly discuss the factors affecting the enzyme action.

3. Clearly define the following terms.
   (a) apoenzyme
   (b) holoenzyme
   (c) coenzyme
   (d) cofactor
   (e) prosthetic group

4. What do you understand by the specificity of enzymes? Briefly describe the 'induced fit' model for the enzyme action.

5. What are serine proteases? What kind of reactions do these catalyse?

6. Explain the importance of charge-relay system in chymotrypsin.

7. What are zymogens? How is the zymogen of chymotrypsin activated?

8. Explain the mechanism of action of α-chymotrypsin.

9. What are cofactors? In what ways do the cofactors help the enzymes in performing their functions?

10. Discuss some synthetic applications of enzymes with the help of examples.

Suggested Readings

Organic Chemistry; I.L.Finar Fifth edition, ELBS and Longman group ltd. (1974).

Chemistry and Biochemistry of Amino Acids Ed. G.C.Barrett Chapman and Hall (1985).

Proteins: Structure and Molecular properties; Thomas E Creighton ; Second Edition W.H.Freeman and company (1993).

Bioorganic chemistry; Herman Dugas; Springer (1999).

GLOSSARY

A

absolute configuration (1.4.1): the actual arrangement of atoms or groups in a molecule, assigned with respect to D - and L - glyceraldehyde molecule.

acidolysis (2.4.5): a reaction involving hydrolysis of a molecule with the help of an acid.

acidosis (4.3.2): an abnormal physiological condition in which the pH of blood is below normal.

acidic proteins (3.3.2): proteins containing a number of amino acids like aspartic acid and glutamic acid with acidic side chains. These have their isoelectric pH in the acidic range.

activating groups (2.4): the chemical groups used to activate a functional group to increase its reactivity in a reaction. For example, the carboxyl group of the N-protected amino acid is activated by converting into its O-succinimide ester or mixed anhydride before coupling it to the second amino acid.

active site (4.3.1): a part at the surface of the enzyme where the substrate molecule binds to the enzyme. It is quite small as compared to the overall size of the enzyme and contains binding sites and catalytic sites.

affinity chromatography (3.4.1): a kind of column chromatography in which the column material has special affinity for the molecule being separated.

albumins (3.2.1): globular proteins which are soluble in water and are heat coagulable. These are found in blood plasma, muscle, egg white, milk and other animal substances along with many plant tissues etc. and participate in the transport of many substances like, metal ions, fatty acids etc.

alkalosis (4.3.2): a physiological condition in which the pH of blood is above normal.

amino acids (1.1): the fundamental structural units of proteins containing an amino group as well as a carboxyl group.

amino acid residue (2.2.2): the amino acids, as they exist in the peptides or proteins. These are the remains of the amino acid in the peptide or protein.

amino acid sequence (3.4): the order of linking of amino acids in a peptide or a protein. The amino acid sequence of a protein is referred to as its primary structure.

amino protecting groups (2.4.1): the chemical groups that are use to protect the NH_2 group of the amino acid during peptide or protein synthesis. These acylate the amino group and thereby substantially reduce its basicity and nucleophilicity.

ammonolysis (1.8): introduction of NH_2 group in a substrate by replacement of halogen with the help of ammonia.

analgesia (2.1): relief of pain without loss of consciousness

antibodies (2.1) : the proteins which bind specifically to the antigens. These specific glycoproteins are produced by the cells of the immune system in response to the antigens and provide protection against them.

antigen (3.2.3): a substance that is foreign to the body and is capable of eliciting a specific immune response.

apoenzyme (4.3): the protein part of the enzyme which requires a cofactor to perform its function.

artereosclerosis (3.2.2): a chronic condition in which the walls of arteries get thickened and rigid and are unable to process adequate supply of blood.

asymmetric synthesis (1.8.2): synthesis of a pure optically active compound using chiral catalyst or by asymmetric induction using chiral oxidations.

azlactones (1.8): cyclic organic molecules obtained by the reaction of an N-acyl amino acid with an aromatic aldehyde in presence of acetic anhydride and sodium acetate.

B

basic proteins (3.3.2): the proteins containing a number of amino acids like lysine, arginine etc. with basic side chains. These have their isoelectric pH in the alkaline range.

benzyloxycarbonyl (or carbobenzoxy (Cbz) or Z) group (2.4.1): an amino protecting group used in peptide synthesis.

betaines (1.6.1): trialkyl derivatives of amino acids i.e. internal quaternary alkyl ammonium salts of amino acids. These are zwitterionic in nature.

binding groups (4.3.1): the amino acid side chains in the active site of the enzyme that help in holding the substrate to the enzyme

binding pocket (4.6.2): a part of the enzyme, where the substrate binds for the enzyme to act.

biotin (4.4.5): a coenzyme for enzymes involved in carboxylation reactions.

biuret reaction (3.3.6): a colour test for inferring the presence of a peptide bond. The biuret reagent (alkaline solution of $CuSO_4$) gives a violet or purple coloured complex with a molecule containing at least two peptide bonds.

boc or *t*-boc group (2.4.1): *t*-butoxycarbonyl group used as an amino protecting group in peptide synthesis.

botulism (3.1): an acute paralytic disease —a type of food poisoning produced by the bacterium, *Clostridium botulinum*, present in improperly cooked or preserved food.

bpoc group (2.4.1): 2-(4-biphenylyl)-isopropoxycarbonyl group used as an amino protecting group in peptide synthesis.

C

carbodiimide coupling method (2.4.4): coupling of an N-protected amino acid or peptide with an amino acid ester with the help of dicyclohexylcarbodiimide (DCC) to give a peptide.

carboxyl protecting groups (2.4.2): the chemical groups that are used to protect the COOH group of the amino acid during peptide or protein synthesis.

catalytic groups (4.3.1): the amino acid side chains in the active site of the enzyme that are involved in the catalytic process. For example, the side chain of aspartic acid, histidine and serine are involved in the catalytic action of chymotrypsin.

catalytic triad (4.6.2): a group of three amino acid side chains (Asp-102, His-57 and Ser-195) that are assoiated with the catatytic activity of the enzyme, chymotrypsin.

chromo-proteins (3.2.2): conjugated proteins having a coloured compound as the prosthetic group.

chymotrypsin (4.6): a proteolytic enzyme produced by pancrease. It is associated with hydrolysis of dietry proteins and cleaves the peptide bond next to an aromatic side chain.

chymotrypsinogen (4.6): the inactive precursor of the proteolytic enzyme, chymotrypsin.

co-carboxylase (4.4.5): a coenzyme that works with the enzyme catalysing the transfer of carboxyl group.

coded amino acids (1.3.1): (also called primary protein amino acids or proteinogenic amino acids) the amino acids found in protein and specifically coded for in the process of translation.

cofactors (4.3): the non-protein prosthetic groups that aid the enzyme in performing the catalytic function.

coenzymes (4.3): the non-protein component associated with enzymes. These may be metal ions or complex organic molecules associated by conjugation. These serve as the donor or acceptor of electrons or functional groups in the reaction.

collagen (3.2.1): a type of fibrous protein having three elongated α- helical strands that spiral around each other like the strands of a rope.

column chromatography (3.4.1): a technique for the separation or purification of the constituents of a mixture of compounds.

configuration (2.2.1): a particular spatial arrangement of atoms or groups in a molecule, characteristic for a given stereoisomer.

conjugated (or complex) proteins (3.2.2): the proteins that exist as complexes with non-protein components.

connective tissue (2.1): the tissue involved in the binding together of various parts and organs of the body.

contractile and mobile proteins (3.2.3): the proteins involved in all forms of movements. For example, myosin and actin are responsible for muscle and heart movement.

coupling of amino acids (2.4): the reaction between two suitably protected amino acids (or peptide fragments) or between suitably protected amino acid and peptide fragments to give a peptide.

charged polar amino acids (1.3.1.1): a class of amino acids having polar side chains due to the presence of positive or negative charge at the end of their side chains.

circular dichroism (CD) (1.4.3): absorption of left and right circularly polarised beams of light to different extents on interaction with chiral molecules.

coupling (2.4.4): generation of a peptide bond from the N- protected and C- activated amino acids during peptide synthesis.

coupling catalyst (2.4.4): catalyst used during coupling step in peptide synthesis to minimise the side reactions. For example, 1- hydroxybenzotriazole (HOBt) or N-hydroxysuccinimide (HOSu)

coupling reagent (2.4.4): a compound used during coupling of amino acids, usually in an inert solvent like methylene chloride or THF, that facilitates dehydration during peptide synthesis

D

Dalton (3.2.2): a unit of mass defined to be equal to 1u (amu).

decarboxylation (1.6.2): removal of a molecule of carbon dioxide from a carboxylic compound e.g., an amino acid.

denaturation (3.3.5): disruption of the overall native structure of a protein on subjecting to heat, x-rays, UV rays, acids, heavy metal ions etc., resulting into its nonfunctional and disorganised form.

deprotection (2.4): removal of the protecting groups from the protected peptide to obtain the free peptide or for getting a free amino or carboxylic group in order to extend the peptide chain.

derived proteins (3.2.2): the degradation products obtained on subjecting the native proteins to different physical or chemical agents.

dielectric constant (3.3.3): a physical property by virtue of which a solvent insulates opposite charges from each other. Ionic or polar molecules have better solubility in a solvent with high dielectric constant.

differential precipitation (3.4.1): a protein separation technique based on differential solubility of the proteins depending on the salt concentration.

differential centrifugation (3.4.1): a method for separating proteins on the basis of their masses. Proteins of different masses separate at different speeds of the centrifuge.

dipeptide (2.2.2): a compound formed by the combination of two amino acids through a peptide bond.

dipole moment (2.2.1): a physical property, defined as the product of charge in electrostatic units (esu) and distance (in cm) separating the atoms in a molecule. It is measured in Debye units, 1 Debye = 3.33564×10^{-3} C m.

$$\mu = e \times d$$

E

Edman degradation (3.4.1.2): a degradation procedure used for sequential determination of N-terminal residues of a peptide.

elastin (3.2.1): a fibrous protein closely related to collagen and constituent of ligaments and walls of the blood vessels etc.

electrophoresis (1.5.1.1): a technique used for separating charged molecules, based on their different mobilities in an electric field.

endocrine (2.1): related to internal secretion by ductless glands

endopeptidases (3.4.1): proteolytic enzymes that cleave the internal peptide bonds of a peptide or protein.

enzymes (3.2.3): the proteins that act as biochemical catalysts.

erythrocytes(2.5.2): red blood cells (RBCs) that carry oxygen to different parts of the body.

essential amino acids (1.3.4): the coded amino acids not synthesised by the body and have to be supplied by diet.

esterification (1.6.2): reaction of an alcohol with an acid to produce ester and water facilitated by using catalysts like conc. H_2SO_4, $C_6H_5SO_3H$. Amino acids also undergo this reaction.

enantiomeric excess (1.8.2): a number representing the formation of an enantiomer in excess as compared to the other during asymmetric synthesis. Also called enantiomeric purity, or percent optical purity and is given as:

$$\% \, ee = \frac{R-\text{enantiomer} \; - \; S-\text{enantiomer}}{R-\text{enantiomer} \; + \; S-\text{enantiomer}} \times 100$$

epimerisation (4.4.1): process of conversion of one epimer into another. Epimers differ from each other in terms of their configuration at only one chiral carbon in a compound containing more than one chiral casbon atoms.

exopeptidases (3.4.1): proteolytic enzymes that cleave the terminal peptide bonds of a peptide or protein.

F

fermentation (1.7): an enzymatically controlled transformation of an organic compound, e.g., conversion of sucrose into glucose and fructose by invertase (an enzyme in yeast). Also used in industrial preparation of α-amino acids.

fibrin (3.2.1): a protein involved in coagulation of blood.

fibrous proteins (3.2.1): proteins having elongated shape like that of fibers or threads. These have high axial ratio.

flavin mononucleotide (FMN) (4.4.3): a coenzyme derived from riboflavin (vitamin B_{12}). It aids certain enzymes involved in redox reactions.

flavin adenine dinucleotide (FAD) (4.4.3): a coenzyme derived from riboflavin, and associated with oxidation-reduction reactions.

flavoproteins (4.4.3): enzymes that require FMN or FAD as cofactors.

fluorescamine reaction (3.3.6): a test used for the detection of amino acids. Fluorescamine gives fluorescent products with amino acids.

fmoc group (2.4.1): 9-fluorenylmethyloxycarbonyl group used as an amino protecting group in peptide synthesis

fragment condensation (2.4.4): a procedure involving condensation of small peptide fragments to make a longer peptide.

full saturation (3.2.2): a protein separation method that involves the coagulation of protein from a solution by saturating it with salt like, ammonium sulphate.

G

gel electrophoresis (3.4.1): a technique for the separation or purification of proteins, based on the differential electrophoretic mobilities of different proteins.

globular proteins (3.2.1): also called spheroproteins or corpuscular proteins, having completely folded polypeptide chains giving a shape of spheroids. These have low axial ratio.

globulins (3.2.1): a type of simple proteins that are insoluble in water but dissolve in dilute salt solution for example, myosin.

glutelins (3.2.1): a type of simple proteins found in wheat flour. It gives cohesiveness to the dough.

glycoproteins (3.2.2): a type of complex or conjugated proteins containing a carbohydrate molecule as the prosthetic group.

glycosidic linkage (3.2.2): a bond that links two monosachharides .

glutathione (2.5.1): a tripeptide found in nearly all the cells of plants, animals and microorganisms. It plays an important role in the biological oxidation-reduction and as a coenzyme.

H

hand and glove model (4.3.1): a model proposed by Koshland Jr. to explain the specificity of enzyme action, the model considers the flexibility of active site of the enzyme.

α-helix (3.4.2) : a sprial conformation for proteins first proposed by the noble laureate Linus Pauling and Robert Corey and is the best known and recognised secondary structure of proteins.

helical wheel (3.4.2): a representation of the amino acid sequence of a helical segment of a protein.

histones (3.2.2): proteins which are rich in basic amino acids like lysine and arginine and are found in eukaryotic chromatin.

HIV protease (2.3.1): a proteolytic enzyme acting on HIV protein.

holoenzyme (4.3): the catalytically active apoenzyme–cofactor complex that binds to the substrate to catalyse the reaction.

homopolymer (2.4.4): a polymer containing same monomeric molecules.

hormone (2.1): a chemical substance, secreted by endocrine glands, having regulatory or stimulatory effect on the activity of cells, tissues and organs, usually located away from point of their origin.

hplc (3.4.1): a type of column chromatography in which the mobile phase is a liquid often under pressure

hyaluronic acid (3.2.2): a simple mucopolysachharide present in the vitreous humor of the eye, umbilical cord and loose connective lissne. It works as a structural element and as a lubricant.

hydantion (1.6.3): organic molecules prepared by reaction of amino acids with phenylisothiocyanate.

hydrazinolysis (3.4.1): a commonly used method for the determination of C-terminal residue of a peptide.

hydrolases (4.2.2): enzymes that catalyse the hydrolytic cleavage of C–O, C–N, C–C and some other bonds.

hyperthermia (4.3.2): a physiological condition in which the body temperature is more than normal.

hypothalamic (2.5.1): related to hypothalamus–a part of brain that lies above the pituitary gland and regulates a number of body functions.

hypothermia (4.3.2): a physiological condition in which the body temperature is less than normal.

I

immune system (2.5.2): an organised group of cells that defend the body against bacterial, viral, parasitic and fungal infection by producing an immune response against the foreign substance.

immunity (2.5.2): capability of body to defend against infections.

immunological (2. 1): related to all aspects of immunity, allergy and hypersensitivity.

inert solvent (2.4.4): generally a non-polar and unreactive liquid acting as a solvent.

insulin (2.5.3): a pancreatic peptide hormone secreted by β-cells of the *islets* of *langerhans* (an endocrine gland). It is associated with metabolism of carbohydrates and maintenance of glucose levels in the blood.

invertase (4.5): an enzyme present in yeast and catalyses the hydrolysis of sucrose into glucose and fructose.

ion exchange chromatography (2.4.5): a separation technique based on the partition of ionic species between a liquid mobile phase and solid polymeric ion exchanger.

IR spectroscopy (1.5.3.4): a spectroscopic technique based on the principle of absorbance of electromagnetic light by molecular vibrations in the IR region of the spectrum and used to decipher the molecular structure.

islets of langerhans (2.5.3): an endocrine gland in the pancreas. It secretes a hormone called insulin that is involved in the metabolism of carbohydrates and controlling blood glucose levels.

isoelectric point or isoelectric pH (pI) (1.5.2): the pH at which the number of positive and negative charges on an amino acid or protein are equal. The molecules at their pI have no net charge and thereby do not move under the influence of electrical field.

isoelectric precipitation (3.3.3): a procedure that employs adjustment of the solution pH to achieve selective precipitation of a given protein.

isomerases (4.2.2): enzymes that catalyse the conversion of a molecule to its isomeric form.

isozymes (4.3): a set of enzymes that act on same substrate and produce same product.

IUBMB (4.2.2): International Union of Biochemistry and Molecular Biology.

K

α-keratins (3.2.1): a type of fibrous protein found in skin, wool, constituted of a number of α-helices twisted together like strands of rope.

β- keratins (3.2.1): a type of fibrous protein found in silk fibroin (a protein obtained from the cocoon of silk moth) and having β-pleated sheet structure.

L

lactam - lactim tautomerism (1.6.3): a special type of tautomerism shown by cyclic amides called lactams.

ligases (4.2.2): enzymes catalysing the joining of two molecules coupled with the hydrolysis of a diphosphate bond in ATP or a similar high energy molecule.

lipids (2.5.2): a class of organic compounds which are insoluble in water but dissolve in a non-polar solvent like ether.

lipoproteins (3.2.2): a type of conjugated protein having lipids as their prosthetic group.

lock and key model (4.3.1): a model proposed by Emil Fischer to explain the specificity of enzyme action.

lyases (4.2.2): enzymes that catalyse cleavage of C–C,C–O,C–N and other bonds by elimination and generate a double bond.

lymphocytes (2.5.2): a type of white blood cells developing in lymphatic tissue and associated with the body's immune system.These constitute about 20 to 30% of the white blood cells of normal human blood.

lysozyme (3.2.1): a bactericidal enzyme found in egg white and human tears, saliva and sweat. It catalyses the hydrolysis of peptidoglycans.

M

mass spectrometry (1.5.3.1): a technique used for structure elucidation of organic molecules based on the principle of generating ions and separating them in a magnetic field on the basis of their mass/charge ratio.

Merrifield synthesis (2.4.5) (solid phase peptide synthesis): a method of synthesising peptides using an insoluble polymeric resin as a solid support.

metal-activated enzymes (4.3): enzymes that require metal ions for their activity. These get activated for their function in presence of the metal ions.

metalloflavoproteins (4.4.3): flavoproteins containing metal ions.

metalloenzymes (4.3): the enzymes with very high affinity for the metal ions. These bind and retain their metal atom(s) under all conditions.

metalloproteins (3.2.2): a type of conjugated or complex protein.

methemoglobin (2.5.2): a compound formed from haemoglobin as a result of oxidation of ferrous to ferric ions.

Millon's test (3.3.6): a test used for detection of proteins. It is based on reaction of the reagent (mercuric sulphate in sulphuric acid) with the amino acid tyrosine.

N

neurological (2.1): related to the diagnosis and treatment of disorders affecting the nervous system.

neuron (2.5.1): nerve cell and any of the conducting cells of the nervous system

neutral polar amino acids (1.3.1): the amino acids having side chains with an affinity for water and are not charged.

nicotinamide adenine dinucleotide (NAD⁺) (4.4.1): a coenzyme derived from niacin (nicotinic acid and nicotinamide) or vitamin B$_3$. It helps the enzymes catalysing oxidation-reduction reactions.

nicotinamide adenine dinucleotide phosphate (NADP⁺) (4.4.1): a coenzyme derived from niacin or vitamin B$_3$. It is associated with oxidoreductases.

ninhydrin reaction (3.3.6): a reaction of amino group of an amino acid or a peptide to give a coloured product, used in the identification of amino acids and peptides.

non-coded amino acids (1.3.3): amino acids not found in protein main chains because of either lack of a specific codon triplet or not arising from protein amino acids and are called non-protein amino acids.

non-essential amino acids (1.3.4): the amino acids which can be synthesised in the body itself and are not necessarily to be taken from the diet, however, these are required by the body as much as the essential amino acids.

non-polar amino acids (1.3.1.1): the amino acids with hydrophobic side chains and therefore usually located in the interior of proteins.

nps group (2.4.1): nitrophenylsulphenyl group used as an amino protecting group in peptide synthesis.

nuclear magnetic resonance (NMR) spectroscopy (1.5.3.2): a spectroscopic technique for structure elucidation of molecules, based on the principle of absorption of ratio frequency radiation by certain nuclei when placed in a magnetic field e.g. ^{1}H NMR and ^{13}C NMR spectra etc.

nucleoproteins (3.2.2): a type of complex or conjugated protein found in the nucleus of a cell.

O

one letter symbol (1.2.1): single letter symbols or codes used for representing amino acid residues in a peptide or protein sequence.

oligopeptide (2.2.2): a peptide containing 3-10 amino acid residues.

orthogonal groups (2.4.1): two protecting groups that do not interfere with the removal of each other

oxytocin (2.4.4): a peptide hormone secreted by pituitary gland and involved in birth and lactation.

oxido-reductases (4.2.2): a class of enzymes catalysing redox reactions.

P

partition chromatography (2.4.5): a separation method including liquid-liquid chromatography in which stationary phase is absorbed on the surface of the column packing.

phthalyl group (2.4.1): an amino protecting group used in peptide synthesis.

peptides (2.1): biological molecules obtained by the condensation of two or more α-amino acids by elimination of water molecules. The amide linkage between two amino acids is called a peptide bond.

peptide bond: (2.1): an amide linkage between two amino acids in a peptide or polypeptide.

pituitary gland (2.5.1): an organ situated just beneath the base of brain and produces a number of hormones.

polymerisation (2.3): a chemical reaction in which two or more molecules combine to form larger molecules called polymers, that contain repeating structural unit.

polypeptide (2.2.2): a peptide containing more than ten amino acid residues.

prolamines (3.2.2): a type of simple proteins that are soluble in alcohol and found in grass seeds.

prosthetic group (3.2.2): the non-protein part of conjugated protein complex.

protamines (3.2.2): a type of simple proteins. These are low molecular weight basic proteins, rich in the amino acid arginine, found associated with DNA in the sperm cells of various animals like fish.

proteases (4.6): a group of enzymes involved in the lydrolysis of peptide bonds of a protein.

protein conformation (3.4.2): the three dimensional arrangement of atoms in a protein molecule obtained by rotation around different bonds without breaking them and stabilised by different types of interactions.

phosphoprotein (3.2.2): a type of complex or conjugated protein containing phosphoric acid as the prosthetic group.

β-pleated sheet (3.4.2): a secondary structure of polypeptides or proteins in which the molecules are aligned side by side in a sheet like arrangement. The sheet is not flat but is slightly pleated.

post-translational modification (1.3): the modification in the amino acid residues of proteins after the synthesis of protein has taken place by the process of translation.

primary protein amino acids (1.1): see coded amino acids.

primary structure of proteins (3.4.1): is concerned with the covalent structure of proteins and refers to the sequence of amino acid residues in the protein.

protecting group (2.4): a group used to selectively derivatise a functional group in a molecule so as to render it inaccessible for the reaction.

proteinogenic amino acids (1.1): see coded amino acids.

proteoses, proteases, proteolytic enzymes (3.4.1): enzymes that hydrolyse polypeptides at specific sites

protic solvent (2.4.2): solvent molecule with a hydrogen atom attached to strongly electronegative element e.g., H_2O.

pyridoxal-5-phosphate (4.4.6): a versatile coenzyme engaged in the catalysis of a variety of reactions in biological system.

pyrolysis (1.6.3): the decomposition of chemical compounds by subjecting them to a very high temperature.

Q

quarternary structure of proteins (3.4.2): an aggregate of two or more polypeptide chains giving biologically functional form of a protein. For example, in haemoglobin four polypeptide chains along with haeme moiety are present in the functional form.

R

racemisation (2.4.2): a reaction in which an optically active compound changes into optically inactive mixture called a racemic mixture. It generally results when chiral molecules are converted into achiral intermediates during a reaction.

random coil (3.4.2): a linear flexible region of polypeptide that has no ordered secondary structure. The denatured proteins also have this conformation.

Ramachandran plot (3.4.2): a plot constituting all possible backbone configurations for an amino acid in a polypeptide.

reductive amination (1.8): reduction of carbonyl compounds in presence of ammonia with the help of a catalyst.

resin (2.4.5): amorphous organic compounds secreted by some plants and insects. These are usually insoluble in water but soluble in organic solvents.

resonance (2.2.1): the property in which a molecule gets stabilised by existing into two or more Lewis structures that contribute to the real structure.

ribosomes (3.2.2): large complexes of proteins and RNA , involved in the synthesis of proteins from mRNA during translation.

ribozymes (4.1): a type of RNA molecules that act as enzymes. These catalyse reactions on other RNA molecules.

Ruhemann's purple (3.4.1): an intense purple coloured product formed by the reaction of ninhydrin with α- amino acids.

S

salting in (3.3.4): a process of increasing the solubility of protein by adding mineral salts to its solution.

salting out (3.3.4): a process of separation of protein from its solution by adding a high concentration of mineral salts.

Sanger (or FNDB) method (3.4.1): a procedure for determining N-terminal residue in peptides or proteins.

Schiff's base (4.4.6): a condensation product of aldehydes and ketones with primary amines.

Schotten Baumen reaction (2.4.1): reaction of acid chlorides with an amine in presence of aqueous NaOH to give amides.

scleroproteins (3.2.2): simple fibrous proteins of animal origin found in connective and skeletal tissues.

secondary protein amino acids (1.3.2): amino acids other than 20 coded amino acids. Sometime in a modification two amino acids may undergo cross linking to form tertiary protein amino acids.

secondary structure of proteins (3.4.2): the local conformation adopted by a polypeptide or protein backbone. α-helix, β-sheet and turns are common secondary structures of the proteins.

semi-essential amino acids (1.3.4): amino acids that are synthesised in the body but not required for the normal growth of the organism. For example, arginine and histidine.

sequenator (3.4.1): a devise that automates Edman degradation procedure for the sequential determination of the peptide sequence.

serine protease (4.6): a protease that uses serine side chain as a nucleophile in its calalytic activity

simple proteins (3.2.2): the proteins composed only of amino acids.

size-exclusion chromatography (2.4.5): a chromatographic separation method that is based on the difference in the sizes of the molecules.

solid phase peptide synthesis (2.4.5): see Merrifield synthesis.

β-strand (3.4.2): a fully extended polypeptide chain which gives rise to the secondary β-sheet structure.

substrate (4.2): a reactant in an enzyme catalysed reaction.

T

C-terminal (2.3): the residue in a peptide or protein that has a free (or derivatised as ester or an amide but not acylating other amino acid) carboxyl group.

N-terminal (2.3): the residue in a peptide or protein that has a free (or acylated but not by an amino acid) amino group.

tertiary protein amino acids (1.3.2): see secondary amino acids.

tertiary structure of protein (3.4.2): the overall three dimensional structure adopted by a polypeptide or protein chain. It consists of different secondary structures stabilised by a number of interactions.

thermocoagulable (3.2.2): that can be coagulated by heat.

three-letter symbol (1.2.1): a method of representing amino acids using first three letters of the common names of the amino acid. The first letter is written in capital and the rest two in small letters.

thiamine pyrophosphate (TPP) (4.4.4): a coenzyme obtained by phosphorylation of thiamin or vitamin B_1.

thiocyanate method (3.4.1): a procedure used for determining the C- terminal amino acid in a peptide or protein.

tendons (3.1): tough fibrous connective tissues by which the muscles are joined to the bone and transmit the force exerted by the muscle.

torsional angles (2.2.1): angles of rotation around different C–C and C–N bonds along the peptide backbone in a peptide or protein. These are useful in describing the conformations of peptides and proteins.

toxins (2.1): a substance produced by micro-organisms causing infection and disease in humans.

transaminases (4.7.6): an important class of enzymes catalyzing the transfer of α-amino group from an α- amino acid to a α -keto acid to give a new amino acid.

transcarboxylation (4.7.5): transfer of a carboxyl group from one species to the other.

transferases (4.2.2): a class of enzymes engaged in catalyzing reactions involving transfer of a group.

transition state (4.3.1): the high energy state that must be achieved by reacting molecules to get converted into the products.

translation (1.3): a stage in protein biosynthesis at the ribosomes in which the coded amino acids are incorporated one by one as per instructions contained in messanger RNA (mRNA) which has been transcribed from the 'gene' for the protein in question.

trityl group (2.4.1): triphenylmethyl group used as an amino protecting group in peptide synthesis.

turnover number (4.3.1): number of molecules of substrate upon which a given molecule of the enzyme acts per second.

U

UDP-glucose (4.4.2): a glucosyl donor for the biosynthesis of glycogen.

urease (4.2): an enzyme acting on urea and decomposing it into CO_2 and H_2O.

ultraviolet spectroscopy (1.5.3.3): a technique used for chemical analysis and structure elucidation, based on the principle of absorption of electromagnetic radiation in the visible or ultraviolet region by molecules showing electronic transition absorbing at characteristic wavelengths.

V

vaccines (2.1): the preparations containing killed microorganisms, living attenuated organisms or fully virulent living organisms. These are administered to produce or increase immunity to a particular disease.

X

xanthoprotic test (3.3.6): a color reaction used for detecting the presence of amino acids like tyrosine or tryptophan in a protein molecule.

Z

zwitterions (1.5.2): the charge separated form of an amino acid that results from the transfer of a proton from a carboxyl group to the amino group, also called a dipolar ion and is represented as.

$$H_3\overset{+}{N}-\underset{R}{\overset{H}{C}}-COO^-$$

zymase (4.5): an enzyme present in yeast and converts glucose and fructose into ethyl alcohol.

zymogen (4.6): the inactive form of a proteolytic enzyme.

Index

A

Acceptor grouptransferase (or donor grouptransferase) 169
Acetoin 201
Acetonide 212
Acetylation 11
Acetyl-CoA carboxylase 202, 203
Achiral 58
Acid chloride 33, 34
Acid hydrolysis 160
Acidosis 99, 181
Acinetobacter 213
Activated carbon 40
Activation 94
Active site 176, 177, 178, 179, 181, 188, 191
Acylation 68, 72, 78
Acyl-enzyme ester 182
Adenosinetriphosphatase 170
Affinity labelling 185
Agglutinin 156
Aggregates 122, 123, 158
Aggregation 125, 126
Albumin 116
Aggregation 101
Alcohol dehydrogenase 165
Aldimine 206, 207
Aldolase 170
Aldolisation 195
Alfalfa 14
Alkaline hydrolysis 39, 132
Alkaline phosphatase 180
Alkalosis 181
Alkylation 40

Alpha amylase 180
Alzheimer's disease 107
Amidases 172
Amino acid analyzer 133, 134
Amino acid residue 5, 7, 12
ε-amino group 6, 10, 12, 137, 144, 207
α-amino orthoesters 48
α-aminonitriles 44
Amino acid residues 122, 123, 126, 131, 132, 134, 141, 143, 144, 148,
Amino acids
    destruction during protein hydrolysis 132
    general formula 2
    C-terminal 5
    N-terminal 5, 12
Amino acid sequence 126, 134, 137, 144, 152, 157
Aminotransferases 206
Ammonium sulphate 116, 117, 120
6 aminopenicillanic acid 216
Ammonolysis 40, 42
Amphipathic helix 153
Amphoteric 22
Amphoteric character 123, 160
Anaerobic 132
Analogues 40
Anchoring 96
Angiotensin II 146, 161
Antibodies 66, 121
Anticoagulant 119
Antigen 121
Antioxidant 62
Antiparallel β-sheet 154
Apoenzyme 190, 219

E

F

G

V

W

X

Y

Z